Palgrave Studies in Cultural Participation

Series Editors
Andrew Miles
Department of Sociology
University of Manchester
Manchester, UK

Lisanne Gibson
University of Dundee
Dundee, UK

This series will provide a platform for contributions to a newly defined field of 'participation studies' (Miles and Gibson, forthcoming 2021). Participation in cultural activities is a research subject within a number of disciplines and fields, ranging from sociology to cultural studies, incorporating tourism, leisure heritage, museum, media, theatre, and cultural policy, to business and management studies. This series will bring together debates across these disciplines to consider the subject of cultural participation in all its dimensions.

The series brings together research on traditional cultural tastes and practices with research on informal 'everyday' activities. In doing so it broadens our understanding of cultural participation, focusing on participation as a pluralistic concern, exploring the links between the cultural, civic and social dimensions of participation, and reconsidering its framing in time and space by political economy, material resource and cultural governance.

Abigail Gilmore

Culture, Participation and Policy in the Municipal Public Park

palgrave
macmillan

Abigail Gilmore
University of Manchester
Manchester, UK

ISSN 2661-8699 ISSN 2661-8702 (electronic)
Palgrave Studies in Cultural Participation
ISBN 978-3-031-44276-6 ISBN 978-3-031-44277-3 (eBook)
https://doi.org/10.1007/978-3-031-44277-3

This Palgrave Macmillan imprint is published by the registered company Springer Nature Switzerland AG.
The registered company address is: Gewerbestrasse 11, 6330 Cham, Switzerland

Paper in this product is recyclable.

This book is dedicated to Juno, my old lady Patterdale terrier, and former accomplice to so many valuable walks in the park.

Acknowledgements

I would like to acknowledge and thank all funders and supporters of the research underpinning this book: Arts & Humanities Research Council (AHRC) Connected Communities: Communities, Culture and Creative Economies and Creative Scotland *Understanding Everyday Participation— Articulating Cultural Values* (grant reference (AH/J005401/1) and AHRC Connected Communities Festival 2015 '*Connecting the museum and the park—everyday participation and community stewardship of local cultural assets*'; University of Manchester School of Arts Languages and Cultures Impact Fund, Social Responsibility Fund and Supplementary Research Leave. With grateful thanks also to the Faberllull Foundation who provided valuable thinking space at the Future of Cultural Institutions residency, Olot, Catalonia in 2019. Heartfelt thanks and warm wishes go to all my fellow researchers on the UEP team, led by Professor Andrew Miles: Professor Ele Belfiore, Dr. Varina Delrieu, Dr. Patrick Doyle, Dr. Jill Ebrey, Dr. Delyth Edwards, Professor Lisanne Gibson, Sarah Hughes, Dr. Felicity James, Dr. Luciana Lang, Dr. Adrian Leguina, Professor Jane Milling, Dr. Susan Oman, Professor Kerrie Schaefer, Dr. Louise Senior, Dr. Mark Taylor, Dr. Ruth Webber, Claire Huyton, Dr. Charlene Linton, Dr. Patrick Doyle, Dr. Ben Dunn and all the coders and other research assistants who worked on the project. UEP turned out to be a very fecund project: a most unified endeavour in radical re-evaluation of cultural practice and value in everyday life, even after we realised how hilariously over-specified and under-resourced it was. Thank you all for your commitment and perseverance in this impactful and important project.

I would also like to thank all participants in the research, and to highlight those from cultural, civic and policy organisations who generously
gave their precious time, including Fran Hayfron, Esme Ward, Helen
Marks, Jo Beggs, Alistair Hudson (The Whitworth/Manchester Museum
Partnership), Amanda Wallace and Hannah Williamson (Manchester Art
Gallery), Emma Anderson and all at Macclesfield Museums. Thank you
for the kind permission to use and reproduce images from your archives.
Huge thanks and kind regards to Dr. Ana Sanchez-Santana who under-
took the PhD researcher-in-residence under my supervision and brought
political sciences acumen to this brief foray into the museum in the park so
nobly. Many thanks go to Kerenza McClarnan, director of Buddleia,
Luciana Lang, Jeni Allison, Torange Khonsari, and Max Dunbar at the
Manchester Jewish Museum for our attempts at commoning in Cheetham
Park, and to the Platt Fields Friends Group, especially Dan Thomson, Jane
Crowley and Richard Stout, to Ruth Colson, Yuxi Wang, Yun Xiong, Liz
Mitchell and Meg Parnell, Platt Hall/Manchester Art Gallery, John
Mouncey and all at Manchester Parks team, for enthusiasm, resilience and
remote working during tough Covid times. Thank you to Saira Qureshi
for permission to use her image of the 'weekly kneel' for Black Lives
Matter. I would also like to thank the group of 'parky academics' whom I
met during lockdown and whose inspiring interdisciplinary work and
chats kept me going in those strange times: Professor Andrew Smith, Dr.
Ian Mell, Dr. Meredith Whitten, Dr. Anna Barker, Dr. Katy Layton-Jones,
Dr. Nicola Dempsey and Ian Baggott-Smith. A massive thanks to Amie
Kirby for her research assistance and diligent proofing.

I would like to thank friends and family: Susan Oman for continually
feeding me internet park nerdery and pulling me out of a Ruskin hole
almost intact, Craig Hughes-Noehrer for being my first reader and Lukas
Hughes-Noehrer for his support, Simon Buckley for rambling and ranting
around Macclesfield Forest. Thank you to my brother, Sebastian Jones,
one-third of my Spanish family for sharing and translating, and my family
in the Northwest of England, Bob, Ava Bea, Ani and Grace, for putting up
with me with only the occasional eyeroll despite very little interest in the
subject matter.

Abigail Gilmore is a senior lecturer, Arts Management and Cultural
Policy, Institute for Cultural Practices, University of Manchester. Her
research concerns participation, policy and place, from a PhD on local
music industries, scenes and policies, to recent research on everyday participation, public space and place governance. Roles include lead for the

Creative Placemaking theme for the Manchester Urban Institute, principal investigator for the *Beyond the Creative City* Manchester-Melbourne-Toronto project and co-investigator for the *Strategic Coordination Hub for Local Policy Innovation Partnerships*, led by Birmingham City-REDI. Recent publications include an edited volume with Victoria Durrer, David Stevenson and Leila Jancovich, 'Cultural Policy Is Local: Understanding Cultural Policy as Situated Practice' (Palgrave Macmillan) and 'Pandemic Culture' with Ben Walmsley and Dave O'Brien (Manchester University Press). She lives in Macclesfield in the North West of England.

Praise for *Culture, Participation and Policy in the Municipal Public Park*

'Finally, we have a book which engages seriously with parks not just as 'recreation' but as a vital part of the social infrastructure and inseparable from fully democratic, locally focused cultural policy. So often overlooked, Abigail Gilmore has pushed public parks directly into full view and provided us with the foundations of a new engagement with these spaces as essential to democratic public participation.'
—Justin O'Connor, Professor of Creative Economy,
University of South Australia

'This book draws on extensive interdisciplinary scholarship for a deep dive into the social and political life of the public park in Britain. From the nineteenth-century creation of the park as an escape from the industrial revolution to the present-day environmental case for them as "the lungs of the city", it offers the park as a metaphor for another way of living as well as a space in which cultural live is lived in all its richness. By examining case studies across the north of England and Scotland, that sit across the publicly managed (by either state or volunteers) to privately owned it considers different approaches to governance and their precariousness in the face of neo-liberalism despite evidence of their public value. The empirical data the book draws on shows how parks are seen as egalitarian and democratic spaces in which we might "practice citizenship". Yet, it also shows the tensions that exist in who decide on their appropriate use. The book makes a strong case for the park as a valuable research site for understanding the importance of as well as the functioning of the cultural commons and also demonstrates the failures of cultural policy to safeguard them as cultural assets. It is a valuable read for anyone interested not only in the public park but in participation and public value, cultural policy and governance.'
—Leila Jancovich, Professor in Cultural Policy and Participation,
University of Leeds, UK

CONTENTS

LIST OF FIGURES

LIST OF TABLES

CHAPTER 1

Everyday Spaces for Participation and Policy

Abstract This book concerns the significance and values of everyday participation in municipal public parks, the connections these have with cultural policy, placemaking and place governance, and to the practising and stewardship of public space. Adopting a critical cultural policy lens, it identifies the municipal public park as a mundane but extraordinarily treasured place for participation and production of cultural values, for regulation, resistance and the practising of citizenship. The first chapter sets out the motivations for studying public parks as spaces for local governance and cultural policy, outlining the book's theoretical orientation and the conceptual terms that it explores and prioritises in subsequent chapters, which concern the values of parks, their 'cultural-ness' and 'public-ness' and their relevance to policy. It also sets out the structure of the book and outlines the empirical research on everyday participation in England and Scotland, the basis for the inquiry into these important cultural policy spaces.

Keywords Municipal park • Cultural policy • Local governance •
Cultural value • Public space • Cultural public sphere • Green lungs •
Public good

© The Author(s), under exclusive license to Springer Nature
Switzerland AG 2023
A. Gilmore, *Culture, Participation and Policy in the Municipal
Public Park*, Palgrave Studies in Cultural Participation ,
https://doi.org/10.1007/978-3-031-44277-3_1

INTRODUCTION

The municipal public park is both a mundane and exceptional object of study. This book explores the significance of urban public parks and the contributions they make in our everyday lives, as communal spaces of participation. I consider them as cultural spaces and resources, component parts of broader cultural ecosystems, in their facility of public space and public value and as long-standing instruments and agents of cultural policy. The impact of the global COVID-19 pandemic throughout 2020 and 2021 significantly heightened this consideration. As coronavirus transmission locked people out of their workplaces, leisure spaces, cultural and night-time economies, it presented the public park in a new light, in an accelerated rise to the top of the essentials list for everyday living, exercise and respite from the torpor of home-schooling, Netflix streaming and Zoom calls.

For those who have had access, and particularly for those without private gardens or means of transport to rural landscapes, parks became green havens offering places to hear birdcalls, breathe fresh air and absorb the natural environment. During national lockdowns, parks were the sites of family parties, birthdays and anniversaries, illicit gatherings, work meetings, daily walks and 'pubstitutes'.[1] On any single evening in early summer 2020, the local English municipal park contained joggers slaloming around picnickers and dog-walkers, children feeding ducks, teenagers hogging playgrounds, knitting groups social distancing on benches, skateboarders dodging mobility scooters, alongside circus skills, personal training and yoga groups. Serendipity and DIY culture augmented existing natural heritage, public art and statuary within parks. Parks became impromptu festival sites, theatres, art galleries and living rooms, imbued with a tacit recognition of the rules of the pandemic – social distancing, family bubbles – and littered with the detritus of mask wearing and hand sanitising amongst the barbecues and beer cans.

> Manchester's Park Team and more than 100 voluntary groups that they support have worked tirelessly to provide safe access to the green lungs of the city. The number of people visiting parks has risen by more than 20% and the pressure to keep them clean and safe has risen accordingly, with 147,000 bags of litter collected from parks last year. (Manchester City Council 2021, p. 130)

The use of parks as cultural spaces, sites of domestic practice and everyday participation, is far from new, nor is the regulation of human behaviour within these spaces, whether through external supervision or internalised convention. As this book argues, municipal public parks support multiple forms of participation that generate cultural value whilst facilitating the regulation and control of individual (and public) bodies and their behaviour. They have done so since their establishment in the nineteenth century, drawing on much older practices of land use and governance within human settlements. As public spaces, they present opportunities for social encounter, observation and judgement that are at once a means for management and a source of public good. The pandemic has made these attributes keener to the public eye, if not more demanding of the attention of those who hold the purse strings. Despite the evidence of its centrality to lockdown life, there is no statutory duty on the part of the government to care for and maintain public parks in England, the main site of this study, and there remain many inequalities of access and diversity of policy support for green spaces in cities all over the world (Shoari et al. 2020).

There are many investigators and supporters of public parks across academic disciplines, from researchers who look at city green spaces from environmental, ecological and urban planning perspectives (e.g. Dobson and Dempsey 2020; Mell 2022) to social scientists who explore parks' relations to social order, public management and economic geography of towns and cities (e.g. Low et al. 2009; Cohen et al. 2016; Barker 2017; Smith 2013). As Jones (2018) argues, parks are a fertile ground for cross-disciplinary study, since they are evolving ecological spaces, which transform and mediate and which can be read through a wide range of lenses. There are park historians who study changes in the aesthetics, dynamics and uses of parks over time (e.g. Colton 2016; Hickman 2013; Layton-Jones 2018; O'Reilly 2019), many of whom write from the perspective of their own public engagement with specific urban parks as curators, voluntary stewards and lay preachers of their value (e.g. Conway 1996; Ruff 2016). There are also those whose focus is on the very concept of the public park as a necessary invention that is both a metaphor for utopia and lost paradise (e.g. Jones and Wills 2005), people's palace and commonplace institution (e.g. Elborough 2016) and a time capsule which offers a text written by the city as "site of serious urban enquiry" (Jones 2018, p. 41). All of these observers are passionate and ardent fans of the park, whatever their disciplinary perspective.

What this book aims to do differently, and the contribution it aims to make to this interdisciplinary terrain, is to explore municipal public parks through a lens of cultural policy to make connections between parks and other forms of local publicly funded, state-supported cultural provision that brings people together in participation in public space, in ways that both satisfy and frustrate policy makers whilst realising a multitude of cultural values. My interest is in how public parks, their histories and uses, are recognised and understood as policy instruments, and also policy problems, within contemporary everyday life. They are, I argue, part of local cultural ecosystems that also include art galleries, museums, libraries and theatres, concert halls and comedy venues, pubs, clubs, community programmes, studio spaces, festivals, gyms, leisure centres, football stadia, church halls and other faith spaces. Their study reveals much to the critical cultural policy researcher and scholar of arts and cultural management, which will complement and contribute to existing knowledge on discourses of cultural value and their association with the regulation and promotion of participation.

It is no accident, therefore, that I begin by turning to a common reference point in cultural policy studies, the English Victorian art critic and social commenter, John Ruskin.

Parks and Value

There is an often-cited John Ruskin quotation that signals the purpose and value through which public parks are esteemed:

> The measure of any great civilisation is its cities and a measure of the city's greatness is to be found in the quality of its public spaces, its parks and its squares. (NHMF and HLF 1996, p. 9)

Lord Rothschild used this quotation in his foreword of the 1995–1996 Annual Report of the Heritage Lottery Fund and National Heritage Memorial Fund, which set out the terms through which the new Urban Parks Programme would invest in public parks to halt their state of rapid deterioration, as visible and popular beneficiaries of the new National Lottery funding. This signalled a specific shift in cultural policy, amongst a number of other fundamental changes during the 1990s, by deliberately extending the breadth of the remit for heritage funders and hence the definition of what comprises heritage value and historic environment (Tandy

2019). It was partly an outcome of *Park Life*, the influential report by consultancy Comedia (Greenhalgh and Worpole 1995), which put forward the recommendation for lottery funding as a necessary means to halt decline, based on substantial empirical research into public perceptions and uses of public parks. This research acknowledged the multifarious historical and contemporaneous attachments of value to parks through their community use and, importantly, highlighted how, despite the changing urban fabric, parks still convey more public value than was recognised by their contemporary place in urban policy agendas. It made an important new case for National Lottery funds to go to parks on the basis that this would meet a significant proportion of the lottery distributors' funding criteria by creating spaces for community wellbeing through sports and recreation, arts and cultural programming, memorialisation and heritage and upkeep of the public realm.

The considerable values that urban parks hold for both people and places, and the articulation of these values as both intrinsic and extrinsic, contributing to a range of policy interests, are central concerns of this book. I will be arguing that the values articulated through the relationships of individual and social bodies with public parks are part of affective communication within the "public cultural sphere" (McGuigan 2004). In addition to social, political, economic and other value domains, parks are vehicles for both imagining the good life and working out salient thoughts and feelings of "life-world concerns" (McGuigan 2004, p. 134), as much as, if not more than, great arts and literature. Through these relationships to the emotional and aesthetic, and their predominantly public ownership, municipal parks are part of urban cultural ecosystems, governed, buffeted and bankrolled by cultural policy flows and forces. This raises a further central theme of the 'parks value paradox': despite an explicit recognition of public value and multiple forms of evidence to support these claims, public funding for municipal public parks in England is of consistent contention and their status as assets, rather than liabilities, for local government in constant jeopardy.

The Ruskin quotation above has frequently been attached to lobbying for greater recognition of the objective value of parks worldwide, appearing in many briefing and advocacy documents, from open letters to newspapers (Barber 1992) and submissions to Select Committee enquiries (House of Commons 1999) to public space strategies in New Zealand (Wellington City Council 2009) and hotel brochures in India (RARE 2020). Its original provenance amongst Ruskin's many writings has been seemingly lost[2]; if genuine, its sentiment originated at a time of continuing concern about the shocks, checks and balances of mass urbanisation and industrialisation in the nineteenth century. It is worth briefly unpacking the statement within this context, to explore how it heralded municipal approaches to cultural policy and why it is important to the arguments of this book.

Within the statement, public squares and parks are components of the urban environment, with qualities which play active parts in making cities great, and great cities, in turn, demonstrate civility: parks, therefore, act as nested quality indicators in the benchmarks for cities and for civil society. To understand better the relations to cultural policy, the logic chain in Ruskin's measure for value needs to be reversed, as in this direction we see that it is the propensity of public parks to instil civilising agendas within urban populations, through their prescription of cultural values for the betterment, enrichment and containment of society. Arguably, parks make the civil population that allows cities to prosper, rather than the other way round. But let us not (yet) get hung up on correlation or causation: a measure is an indicator, not a predictor, presumably.

Further, a number of things can be construed by the term 'quality' as a measure for how parks are related to civilising processes. 'Quality' might mean the design and architectural standards of these new public spaces, the ways that the location, planning and planting of parks present land-scapes and vistas within cities as incursions of nature in the urban realm. It might mean the affordances that their designs have in bringing together different publics in social encounter and interaction, creating zones and amenities that require people to assemble, to observe each other, take part in exercise and keep them away from less healthy pursuits. These were all strategies of the Victorian parkmakers at the time of Ruskin's writing, as is discussed later, in Chap. 2.

Ruskin is known for his art criticism, his philanthropy and his polemics against the political economy of industrial capitalism in his vast collected works of lectures and public letters, published assiduously in *Fors Clavigera*. Whilst he advocated for the promotion of engagement with nature and beauty, he also knew that parks were places of social control and class distinction. For example, he finds parks to be the display sites for wealth and self-aggrandisement (and self-containment) for the un-curious, where Victorian gentlemen would wander "my richer readers, only round the parks, every day, instead of from place to place through England, learning a thing or two on the road?" (Letter 66, June 1876, *Fors Clavigera Volume*

VI in Cook and Wedderburn 1907, p. 631). Here Ruskin is berating his peers for not wanting to see the different walks of life that he accesses as he travels around the country on his own peculiar civic mission. Another earlier letter features a diatribe on London society as a drain on the rest of the country, whose country people provide its food, and whose manufacturing classes provide clothing, "iron railings, vulgar upholstery, jewels, toys, liveries, lace, and other means of dissipation and dishonour of life". In this, he describes the metropolitan classes as "Park Squirrels" and Hyde Park, London, as a "great rotatory form of the vast squirrel-cage; round and round it go the idle company, in their reversed streams, urging themselves to their necessary exercise" (Letter 44, August 1874, p. 136). Ruskin is therefore aware of (and I would argue ambivalent about) the qualities of parks as sites for the promotion and display of *certain* tastes, and values, and the parading of aesthetic judgement, cultural taste and status. These are juxtaposed with his interests in opening up common land for productive use, in creating a 'national store' of treasures and supporting education in the arts through the Guild of St George (Hewison 2018).

As contemporary municipal strategies, parks are also the sites of co-located museums to provide arts education for working people, as proposed by Ruskin contemporary and Manchester Art Museum founder, Thomas Horsfall, in an open letter to the Manchester Guardian in 1977, which was promoted by Ruskin in his own open letter to the working people of Sheffield:

> In each of our parks a small gallery of the kind might be formed, which might of course, also contain a few good engravings, good vases, and good casts, each with a carefully written explanation of our reasons for thinking it good. (Horsfall, cited in Ruskin, Letter 79, Life Guards of New Life, July 1877, *Fors Clavigera Volume VII*, p. 155)

New Victorian parks often held remnants of their previously private status such as grand houses and halls to which an art museum brought new purpose, or if not, a reclaimed manor house site provided the opportunity for new architectural practice. The gallery in the park encapsulated Ruskin's and Horsfall's ambitions to bring together art and nature with moral improvement and education, housed within city limits and conserving collections and assets that could be accessible equally to all, as discussed further in Chap. 3. Its influence can be seen on the policy and practice of museums and their curators in engaging communities from this period until today (Nutter 2020).

This rather Arnoldian strategy to bring culture democratically to the masses is quite distinct from the term 'green lungs', the most often used nomenclature for the function of public parks in the city (Jones 2018). The analogy of parks providing vital organs, allowing the city to breathe and combat and process pollutants, remained remarkably intransigent since it first came into use in the early days of parkmaking and advocacy over 200 years ago.[3] As Crompton (2017) documents, the enduring power of the 'lungs' metaphor lies in its ability to convey the significance of parks for the health of the city as a social and labouring body, off-setting impediments to a working populace, such as airborne disease and poor physical health, whilst obscuring a further dimension to the managed incursion of natural green spaces in the city fabric. This second dimension was the private (and classed) cultural practice of walking in natural land-scapes which emerged as a signifier of elevated taste and distinction in the eighteenth century, levering the interest of the higher social classes in cre-ating spaces for leisure that were safe from the by-products of industrial capitalism, such as overcrowded and unsanitary city streets. These prac-tices were endowed through the metaphor of green lungs to the broader population. The creation of public parks in England followed the 1833 House of Commons Select Committee on Public Walks, a pivotal moment in urban reform that brought bounded *walking space*, rather than parks per se, to the masses, alongside the benefits of greenery:

> Thus, as a "private good" at the individual level, the metaphor suggested that parks were a safe and healthy place to exercise. At the same time, as rare oases of vegetation in the dense industrial urban areas, parks performed a "public good" communitywide function. (Crompton 2017, p. 106)

The environmental benefits claimed for greening the city, promoting biodiversity and supplying oxygen afforded by plants and trees, alongside other significant advances in scientific enquiry and public health policy, are only part of the story of parkmaking. Throughout the municipal public park's history, it is the harnessing of nature in the service of urban plan-ning and social control that speaks loudest to policy, and in particular cultural policy, as this book attests.

Aside from the foregrounding of nature in opposition to culture, there is common ground between the art critic-cum-political commentator and the early Victorian advocate for public health reform when reckoning the value of parks. Both Ruskin's quality metrics and the lungs-of-the-city metaphor position the park as a component part of a larger urban ecology. They suggest that park advocacy enshrines value systems that are based not just on evidence of public benefit but also hierarchies of taste and aspiration and judgement of behaviour in public space, closely tied to socio-economic class, industry (in the Victorian sense) and the labouring body. Both positions recognise the park as a space of potential and agency, influence and reform, as means for practising policy and as sites of practice through which the city and its citizens are made and remade.

In this book, I consider the evidence for the enduring contribution that public parks make in everyday life and attempt to make a case for their valuation as part of broader urban cultural ecosystems, as agents of cultural policy. Like Jones (2018), I argue that parks provide a lens through which to examine the city, the urban fabric and the lives of denizens woven into it. Furthermore, as the property of local authorities and charitable trusts, they are part of the infrastructure for place governance and cultural management, and their study offers new insights into the values and relations of local and national policy and planning. Before going on to say how the book reaches this proposal by outlining its methodology and structure, I want to unpack some key terms and explain a little about the conceptual selections I have made.

Practising Public Space

If, as suggested above, municipal parks provide spaces within urban contexts for practising citizenship and participation in everyday live, how can we understand this role in relation to public space?

In Jan Gehl's 'Life Between Buildings: Using Public Space' (Gehl 2011) first published in 1971, there is a pivotal recognition that everyday life practices that are necessary, functional and requisite, and those that are optional, motivated by desire and interest, are brought together as social activities through the qualities of the public realm. This is because when conditions are optimised – when public space is conducive – more people opt to use it outside of *when they have to*, so the frequency of uses (both necessary and optional) and therefore the chances for social interaction increase. Social connections and interactions happen sporadically and serendipitously because they are visible, observed and hence accessible:

They develop in connection with the other activities because people are in the same space, meet, pass by one another, or are merely within view. (Gehl 2011, p. 12)

Finding ways to improve social contact is clearly important for those who can inform the qualities and properties of public space – the architects, planners and also the parkmakers, although Gehl does qualify this position somewhat, suggesting there is no direct causation between the physical frameworks of public space and the intensity of social connections, their specific nature or outcomes. Rather it is the creation of opportunity and possibility within public spaces that afford their sociality.

Gehl's thesis is explicitly concerned with life *between* buildings, in the sense of proximal public space defined through its relationships to urban physical frameworks, comprising streets and street furniture, squares and their edges, architectural statements and features. It builds on and critiques the progress of functionalist planning which he observed had led to dispersed and single function sites separated by green space, leaving little motivation or opportunity for social contact in public space. The functionalist influence can be seen to affect parallel and later developments prioritising single-family dwellings disconnected by cars, roads and highways that take life around the buildings (rather than between them) and in which privatised, passive consumption and indirect communication are facilitated by domestic technologies of television and internet. Gehl's consternation is that "something is missing" (p. 49) and that something is active participation and co-presence.

Being among others, seeing and hearing others, receiving impulses from others, imply positive experiences, alternatives to being alone. One is not necessarily with a specific person, but one is, nevertheless, with others. As opposed to being a passive observer of other people's experiences on television or video or film, in public spaces the individual himself is present, participating in a modest way, but most definitely participating. (Gehl 2011, p. 17)

This recognition is fundamental in connecting the affordances of public space to their use, whilst suggesting a separation between (the planning of) physical fabric and social practice. It is in a sense an a priori assumption that leads with the proposition that qualities of public space can create more or less social contact, rather than the obverse, where the possibilities of social contact affect the qualities of public space. The qualities proposed by Gehl that confer social interaction start from the perspective of the human body, its sight lines, and senses, as a pedestrian in motion and in stasis, and radiate outwards to include a variety of recommendations for design features that encourage people to dwell in public space, to hang out, and converse. These are both pro-active – interesting things to look at, supports to rest or lean upon in public squares – and responsive to environmental and climatic conditions, such as light, wind and rain, and aim to encourage people to stay longer outside in plain view. They understand and prioritise human limitations, for example the maximum distance from which we can recognise each other and length of time before walking becomes tiring, particularly when the destination is in sight but not in reach through a direct route. Rather than defining precisely the properties required of public space, Gehl outlines the most important constraints to be planned in or designed out in order that life between buildings results in the optimal potential for social activities, encounters and contact.

Gehl's focus is on the 'modest', the proximate and achievable at a micro level, on the forms of public space that present possibilities and create atmosphere conducive to living in the everyday. In contrast, Sennett's meditation on city spaces and urbanism in 'Building the City' begins at a meta level by differentiating between the *ville* and the *cité* (Sennett 2018). These terms derive from distinct meanings attributed to two types of cities, one of Man and one of God, one substantive, physical and material, the other epistemological, comprising perceptions and beliefs. The *ville* represents the planned environment of the city: not simply the structures and infrastructures of its material construction but the politics and intentions they represent. The *cité*, in contrast, is the lived experience of the city: "the feelings people harbour[ed] about neighbours and strangers and attachments to place" (2018, p. 1). The two are intimately connected, but not necessarily harmonious. They rehearse Mouffe's (2005, 2013) articulation of agonism as contrasting spatial and political logics destined to exist in unresolved tension with one another, as conflict not between enemies but between adversaries. Sennett ponders whether the *cité* and the *ville* might be brought together, connected and integrated, through openness to experiment and difference, and through 'modest making', the inclination of ordinary man [*sic*] to make small, quality adjustments to live their lives and promote the lives of others.

Another important distinction Sennett makes is between synchronous and sequential spaces, which inform how activities can be planned into how people come together. The former is represented by the agora, or market, and the latter the pnyx, or theatre. In the first, people can undertake many different activities side by side at the same time, and the second, the crowd focuses on one activity at a time. The agora is where discussion and debate take place, news is heard, gossip shared and behaviour learnt; it is edgier and less safe, but less passive than the theatre. It encourages sociability and encounter. Sennett looks for methods to plan synchronicity into cities, considering how many different activities can take place in public space at any time, and recommending that social mixing is invited, not imposed, by the architectural forms and functions of public space.

To do this, Sennett identifies five 'open' forms important to creating an 'open *ville*', which include those that promote synchronous activity, allowing many things to happen and be co-present in the same and proximate spaces, those that make a feature of porosity, and privilege borders over the boundary; an example of this would be that parks should not have high fences but can signal where they start and end in softer ways, through 'knee rails' and planting. He also suggests the inclusion and creation of punctuation marks within public realm, through monuments and markers which modestly help give places without much character some definition, and signpost routes around spaces, alongside the use of varying architectural 'type-forms' to differentiate buildings, with different periods and styles, and finally, the notion of seed-planning. Seed-planning involves the instigation of multiple forms of activities which can then be allowed to take their characteristics, for example, street markets which evolve differently based on supply and demand, or raised beds and spaces given over for community gardens (Sennett 2018, pp. 205–241).

The public spaces of municipal urban parks have many of these implicit qualities proposed by Gehl and Sennett. By design, they are open, with sight lines, distances and destinations that are manageable, like benches to dwell on and interesting things to look at close up and further off, soft edges, distinct zones, and opportunities for synchronous and sequential activities side by side. As parkmakers and advocates since the very first parks have recognised, they are arenas for social encounter. Likewise, for Barker et al. (2019) they "mediate conviviality" (p. 10), by offering the possibility of short, limited but meaningful contacts that foster park norms and permit the sharing of public space, despite a variety of differences and distinctions between users and uses:

Social interactions are likely to be limited to observation of others with few opportunities for routinely negotiating difference through direct verbal contact … the variegated design of many parks and the tendency for self-segmentation, on the one hand, may limit the capacity of these spaces to provide opportunities for thick forms of engagement and encounter but, on the other hand, are valuable precisely because of this, in that they enable different groups to share space, see each other and form an indifference to difference. (Barker et al. 2019, p. 10)

This proposition goes further than Gehl's life between buildings, identifying social relations and the social glue that holds them together as the focal point. Parks provide the means for what Barker et al. (2019) call a "looser social capital" (p. 11) as relations are held within the fleeting tendency of park encounters; they offer an invitation to social mixing which is not imposed (Sennett 2018). They are means through which to recognise difference and develop tolerance, through visibility and community surveillance, and when well-made and used can enhance a sense of security in public space.

The 'public-ness' of public parks is not simply due to park users' visibility amongst others. The optional offer of the park, in Gehl's terms, situates parks as the kind of public space that elicits participation motivated by desire and interest, rather than compulsion or necessity – although during pandemic this balance may have tipped, as parks became one of the few places that successfully overcame the constraints of lockdowns outside the home (and so became more requirement than choice). The types of participation in public parks are various and significant, regulated by self-restraint and community surveillance, and shaped by social norms and cultural codes. They are mundane and of the everyday, a common experience of many, indeed most, urban dwellers. They are also, as I argue in this book, complex spaces of signification, stratified and powerfully interwoven with personal experiences, memories and feelings, evoking senses of place and sense of belonging to communities, families and faiths.

Participation in public parks offers a means for representation of self within wider constituencies, signalling membership of social collectives even when these opportunities are fleeting. Parks act as arenas – agora and theatre – in which to be identified, but not scrutinised too closely, as denizens of place. Discussing the potential for developing a social psychology of public space, Di Masso considers citizenship as a "locational experience that usually unfolds in public space" (2015, p. 65) which is connected – hinged – via the 'right to the city', following LeFebvre's (1968) original conceptualisation. This is defined as the legitimacy of occupancy, the right of the urban dweller to freely access and use public space and to be recognised as part of "the public" (Di Masso 2015, p. 67). He identifies three perspectives on the ways public spaces relate to the practices of citizenship by virtue of their propensity to inform social integration, as 'emplaced practices'. These question how the occupancy of public space is legitimated and vetted, and who by, and express different approaches to the preservation, equality and struggle over the right to the city.

The first is optimistic, assuming a progressive human tendency to seek out and make new spaces for social encounters that bring together diverse forms of social life, in, for example, shopping malls and the internet. This position proposes a universal right to the city; it expresses hope that even when new technologies encroach on traditional forms, this right prevails. The second position, which Di Masso calls 'terminal', foretells the decline of public space through its commodification and privatisation by the market, which strip traditional spaces for communing of the properties that support social interaction and neighbourliness, and enclose them as gated communities. In this scenario, the public-ness of public space is constantly in depletion, eroded by the encroaching private sphere. By contrast, the third position emphasises the role of conflict and struggle involved in making public space:

> [it] defends the idea that public space *has never really been public*, but rather, it is a terrain that is historically defined by the exclusion of disadvantaged groups that use public space to claim public attention and acceptance as regular citizens (e.g., slaves, women, barbarians, children, immigrants, teenagers, drunks, homeless, sex workers, etc.). From this perspective, urban conflict is a central component of public space and a fundamental instrument for achieving the right to the city because it expresses territorially structural power struggles between the accepted publics and the socially unwanted counter-publics". (original emphasis, Di Masso 2015, p. 70)

According to this position, the making of public space is contingent on what (and who) defines good and bad behaviour. Claims to be within public space are articulated through the 'emplaced practice' of locational citizenship, through "ordinary bodily gestures, spatial uses and common-sense beliefs about 'normal' and 'inappropriate' behaviour in public" (Di

Masso 2015, p. 80). These practices, and the conflicts and micro-politics that define and position them, are mechanisms for self-regulation and governance that make and unmake citizenship, within the public sphere. They are constitutive of public space as space, following Massey, which is never finished and "always under construction" (Massey 2005, p. 9).

I argue here that everyday participation practices in public parks are important means for representing self in the negotiations between individual and state, publics and counter-publics when practising public space. Municipal public parks are arenas for the emplaced practices of locational citizenship and places where the public body can be made, institutionalised, performed, challenged and experienced on a daily basis (Di Masso 2012). Thinking in this way raises the question of whose responsibility it is to regulate and retain public space for such practices. As well as constituting the mechanism to claim one's right to be in the public sphere, the public-ness of public parks also resides in their ownership and management by the state. As we will see, this creates interdependency between the recognition of responsibility to provide and protect these spaces on behalf of the public, and the role of parks in mediating, facilitating and governing emplaced practices of citizenship. Public access and use of parks and claims over them as public spaces therefore lie in tension with the expectation of government to look after these spaces and provide them as resources, constituting a finely balanced agonism, following Mouffe (2005, 2013). As a framework through which to bring together in consideration participation in the everyday, the management of public spaces, the 'right to the park', accountability and responsibility of the state, and the contingency of citizenship on practised space, I look to the concept of 'the commons'.

Sharing Management Through Participation

The commons has gained much traction in recent research as a progressive framework for managing resources, particularly in connection to urban and environmental studies (e.g. Gidwani and Baviskar 2011; Standing 2019). Building on the early mediaeval systems of user rights to grazing, harvesting and foraging on common land in times, its conceptual potential lies in an ambition to promote equal and shared ownership and access to non-economic resources, whilst avoiding extractive relationships. At its very simplest, the claim of the commons is its antithesis of private property rights, which has led to the critique that commons are unmanaged and vulnerable to private and rivalrous behaviour rather than collective interest and investment. The argument is that permissive access leads to plunder and depletion of finite resources, and a "tragedy of the commons", following Hardin (1968).

However, where once this tragedy dispelled the power of open and free access due to overexploitation from unruly commoners, recent understanding suggests that the risk of depletion can be regulated and even reversed, by drawing on the social relationships that comprise the commons. Elinor Ostrom's pivotal intervention on governance of the commons (1990) forms a rebuttal that participation in the commons only ends in its tragic loss, by making the distinction between common-pool property and the commons. Here, there can be individual ownership of property with access granted to others under certain conditions that are commonly agreed, allowing the use of resources to be managed by a community. This emphasis on public compact to manage commons is not just about extending access but is aligned with social justice and environmental sustainability. It provides protection of the natural commons in the public interest and defends it from plunder from the powerful (not the commoner), whether through enclosure of the mediaeval forest and parklands for royal hunting, the systematic taking of open field systems of agricultural land into private ownership under the Enclosure Acts in eighteenth- and nineteenth-century England, the colonisation and over-exploitation of natural resources by the Global North, or the dispossession of common property through neo-liberal housing policies, in what Hodkinson calls 'new urban enclosure' (Hodkinson 2012).

Acknowledging this continuity, Standing (2019) expands the notion of the commons to the domains of social, civil, cultural and knowledge to argue for a new Charter of the Commons (echoing the 800-year old Charter of the Forest) and sets out the defence for a structured and transformative society, through communal sharing of and access to social welfare. The commons as a transformative and resistance space can be seen, on a micro-level, in urban commons of the twenty-first century as communities create cooperative structures for managing shared resources, for example, through allotment schemes, guerrilla gardening and play streets (Mackay 2010; Gidwani and Baviskar 2011). These commons are part of wider urban ecosystems, which determine their physical locations, spatial and cultural characteristics and the challenges that they face to maintain access and protect use rights. They are not only resistant to enclosure but are generative because they rely on entering into a compact through participation; in order that commons exist, people give their agreement to act in the interests of commoning, and through their commoning they produce further resources. These additional resources are extra-material, they are not the land or the grain or the firewood but rather are found within the labour of participation and a sense of shared purpose (Bollier and Helfrich 2015).

Similarly, Linebaugh (2014) identifies a symbiosis between the necessary participation to create the commons, and the commons as a requisite condition for participation. Commoning is a gerund, which creates a paradox, since as a practice which has dynamic and collective properties, it is not neutral and may not include everyone:

> Commoning is exclusive inasmuch as it requires participation. It must be entered into. (Linebaugh 2014, p. 15)

This beggars the question of who and what might be excluded from participation in the commons, and the implications of these exclusions for the practising of public space and an agonistic democratic public sphere. In their volume of essays, 'Undercommons: fugitive planning and black study', Harney and Moten (2013) take issue with the commons for its contingency with enclosure and see commoning not as means but as ends, the consequence of which are a continuing cycle of social reproduction, of post-Fordism, racism and inequality.

> But in the moment of right/s the commons is already gone in the movement to and of the common that surrounds it and its enclosure. What's left is politics but even the politics of the commons, of the resistance to enclosure, can only be a *politics of ends*, a rectitude aimed at the regulatory end of the commons. (my emphasis, Harney and Moten 2013, p. 25)

This challenge to the progressive model of the commons points to its collusion with policy as something already co-opting and enacting, at the very moment when participation within it occurs. Harney and Moten instead prefer outside (below) the commons, within the everyday spaces of the "undercommons". They distinguish between planning and policy, where policy is about correction, and fixing others, applying a theorised corrective upon those who had not necessarily sought correction. In this lexicon, policy is a political ontology that thinks on behalf of others, on the basis that they need fixing and that they are unable to think for themselves (Graham 2023). Like the commons, policy requires participation in action and change, but participation of the right kind, from those who have *already been fixed*. Those who inhabit the undercommons also participate, but before they have been fixed. In this realm, Harney and Moten find:

> the means, which is to say the planners, are still part of the plan. And the plan is to invent the means in a common experiment launched from any kitchen, any back porch, any basement, any hall, any park bench, any improvised party, every night. (p. 74)

To be fixed is to be part of policy and is an exclusive practice. To participate as a planner you need the undercommons: the undercommons becomes a space that can hold agonism as a resource, as a tool to make and practise public space, going back to Di Masso's ideas of emplaced practice. The planners are consciously therefore not part of the commons.

Thinking about public parks in relation to commons *and* the undercommons requires understanding how parks support the practising of policy, and whether they provide means that are part of the plan. In the chapters that follow, I consider commoning and undercommoning through everyday participation in municipal public parks. I explore how common agreements are entered into through participation, and where the plunder of shared resources and enclosure of public and social space occur. I also consider what it might tell us about the governance of public parks and their role within cultural ecosystems and urban cultural policy. As spaces where commoning and public policies meet, what forms of participation are entered into, and how do these resist or reject enclosure? How do they practise and rival public space, and what is their relationship to the practising of *cultural* policy?

THE CULTURAL POLICY-NESS OF PUBLIC PARKS

Before exploring these questions, I want to consider formulations of cultural policy, in case it is not yet clear why we should consider public parks as cultural policy spaces. Scholars of cultural policy have defined the field variously. There are too many contributions to this interdisciplinary area to comprehensively include here, nor does this book's argument rely on a historiography of cultural policy studies; however, a brief survey of definitions is needed.

Miller and Yudice (2002) describe cultural policies as "institutional supports that channel both creative and collective ways of life...[as] systematic regulatory guides to action that are adopted by organisations to achieve their goals" (p. 1). Sticking with the notion of channelling and imposing practice, Pykkonen, Simanainen and Sokka (2009) define cultural policies as "the regularizing aspects of politics that imply the coordination of acts, and measure and regulate" (p. 5). Ahearne and Bennett (2009) take a more multifaceted view on where cultural policies originate and emphasise the more generative aspects of cultural policy as "the promotion or prohibition of cultural practices and values by governments, corporations, other institutions and individuals" (p. 27).

Such dualisms, of regulation and channelling, promotion and prohibition, are common when theorising cultural policy. Ahearne and Bennett (2009) contrast policies that are explicitly cultural, nominal and directly concerned with the state's relation to arts and culture, with those that are indirect and implicit, affecting cultural expression, representation and production, but unnamed or attached to other policy agendas such as economic development, health or education. This echoes a distinction, made within the typology of state/cultural relations first identified by Raymond Williams (1984) between the two categories of cultural policy 'proper' and cultural policy 'as display' which was extenuated by McGuigan (2004, p. 63). For McGuigan, there are five ideal types of relations: public patronage of the arts, media regulation and negotiated construction of cultural identity, which are considered 'proper' cultural policy, and national aggrandisement and economic reduction, which are policy 'as display'. Each evoke an instrumentalisation of culture within state policy for non-cultural ends and highlight the political and ideological means to these ends found lurking amongst welfare state models of cultural policy most common in Europe and in European-influenced nation-states, which expect that culture offers something back to society in return for its state sponsorship (McGuigan 2004).

First, however, there is a need to complicate how culture is defined in order to demonstrate the reasons for distinguishing policies as *cultural*, whether implicit, nominal, proper or otherwise. Raymond Williams is a useful qualifying presence here, bringing together understandings of culture from anthropological, spiritual and material senses, as ways of life that involve flows of relations between and amongst people which carry values and beliefs, with a recognition of the relevance of creative production and cultural expression to representation and identity. He has also pointed out culture as being one of the two or three most difficult concepts, along with 'community', to unpack and settle into (Williams 1988). This is not only due to multiple and sometimes competing definitions of culture. It is also that, through a lens of cultural materialism, the attachment of meanings to representations of identity leads to distinctions, differences and hierarchies of value (and taste) and therefore power. Taking his lead from Williams, McGuigan's long engagement with issues of cultural value is framed by his critical intervention on cultural populism which lays bare the tensions between intellectual claims for and against popular and mass, low and high culture (everyday common culture opposed to culture with a big 'C') (McGuigan 1992), revealing their connections to and interests in political economies of consumerism, globalisation, neoliberalism (McGuigan 2004) and later 'cool capitalism' (McGuigan 2009).

In *Rethinking Cultural Policy*, McGuigan follows a line thrown out by Rifkin (2000) that identifies cultural capitalism, the commodification of experience and the rise of knowledge and network economies as the cause of enclosure of the cultural commons and its plunder by cool capitalists. For example, the case of world music is used as an example by Rifkin to demonstrate "[a] gigantic cultural mining and commodification of difference" (McGuigan 2004, p. 131). This appropriation of culture can be met with resistance primarily through civil society, via third-sector groups, non-profit arts companies and community-led organisations, such as churches, that are part of a broader coalition of culture, and more globally, through anti-capitalist issue-based campaigns and social justice movements. Written before the rise in Western ideologies of 'big society', citizen labour and volunteerism, deracinated by the deepening of austerity, entrenched social division and culture wars, this discussion might seem prescient if not wide-eyed and naïve. However, it brokers the question of the role of critical analysts of cultural policy and cultural production and allows McGuigan to re-present a framework for addressing and re-thinking cultural policy, via the cultural public sphere. Building on Habermas' literary public sphere (1992) and distinct from the political public sphere, this concept is defined as:

> the articulation of politics, public and personal, as a contested terrain through affective – aesthetic and emotional – modes of communication. The cultural public sphere trades in pleasures and pains that are experienced vicariously through willing suspension of disbelief, for instance, in even the most mundane soap operas, identifying with characters and their problems, talking and arguing with our friends and relatives about what they should

and shouldn't do … affective communication helps us make sense of life-world concerns, however diffusely. The cultural public sphere provides us with vehicles for thought and feeling on salient issues that matter for whatever reason. (McGuigan 2004, p. 134)

For McGuigan, public culture, mediated (and I would add *practised*) through the cultural public sphere, is not the same as public-sector culture, which has been largely reduced by instrumentalist cultural policy to a spurious set of economic rationales, leaving the significance of 'cultural' empty-handed in defence of its own plunder. The public cultural sphere is a space to overturn the insidious direction of neo-liberal cultural policy – it is "a space for making sense collectively that is more than making a living or exercising power over others" (2004, p. 141). Like the undercommons, the 'cultural' is returned to significance within the cultural public sphere as a means to move beyond the economic and political and to create agency for social justice that is collective (not corrective) and premised on participation. This understanding of the cultural is also a solution to a problem that Ostrom found within common-pool property management. It echoes the role that she identifies for culture as the communication (and practice) of shared norms, consensual frameworks and narratives which are passed down generationally, which allow good management of the commons and define what is seen as "useful" within common resources (Ward 2014, p. 59). In other words, it provides the means to articulate and realise cultural values.

Both Rifkin's discussion about the plunder of the cultural commons and McGuigan's rethinking of cultural policy, its instrumentalism and complicity in the reproduction of cultural capitalism concern more recognised cultural forms and practices, not necessarily with a big 'C' but certainly those that might be defined as popular, folk and mass culture. One might question where municipal public parks fit into this argument, if they are part of the infrastructure for the cultural public sphere, whether and how they are cultural. In this final section of this introductory chapter, to explain how they do (fit) and are (cultural), I briefly set out the importance of everyday participation to the methodology of this book and outline the book structure.

THE CULTURAL-NESS OF EVERYDAY PARTICIPATION
IN PARKS

I did not choose to look at municipal urban parks as public spaces for locational citizenship and emplaced practice, urban commons, policy instruments or environments for the facilitation of the cultural public sphere. Rather, I would like to think they chose me when they emerged as significant cultural spaces that offered the kinds of public value many publicly funded arts and cultural institutions can only dream of to the participants in a major study, carried out between 2012 and 2018 in England and Scotland. Funded by the Arts & Humanities Research Council, *Understanding Everyday Participation – Articulating Cultural Values*[4] was a mixed methods research project which aimed to radically re-evaluate the meanings, practices and values that form everyday cultural lives. This project was interested in addressing the increasing technocracy through which arts and cultural policies are appraised and by which their impact and effects are evaluated, arguably reproducing formulations of cultural policy which serve elite political interests and increasing the potential for vernacular cultural practices to be obscured and under-valued.

The project's central premise is that orthodox models of culture connected to the broader conceptualisation of the creative economy are based on a narrow definition of participation: one that captures engagement with traditional institutions such as museums and galleries but overlooks more informal activities such as community festivals and hobbies. The project aimed to paint a broader picture of how people make their lives through culture and in particular how communities are formed and connected through participation. It was empirically driven using methods including ethnography, intensive qualitative household interviews, archival research, statistical data mapping and re-analysis, workshops and participatory research, in six 'cultural ecosystem' case study sites selected through a 'straw' matrix framework that identified levels of investment against levels of participation, according to identified indicators.[5] Researchers came from varying arts and humanities disciplines, including sociology, cultural policy studies, drama and classics, and took various roles in leading case study fieldwork, quantitative and qualitative analysis, and policy engagement. My role was to lead the Manchester and Salford ecosystem case study which spanned two local authority areas and involved archival research, data collection and analysis through interviews, ethnography, stakeholder workshops, focus groups and assets data mapping in the wards of Cheetham, North Manchester and Broughton, East Salford between July 2012 and September 2013 (see Gilmore 2017; Gilmore and Doyle 2019). In 2014–2015 I also led a follow-on project with postgraduate researchers and an artist in residence with the Manchester Jewish Museum in Cheetham Park (Gilmore and Lang 2020) and a further community engagement project, in 2019–2020, in Platt Fields Park, with the Platt Hall 'In Between' project, Manchester Art Gallery, Manchester City Council and the Friends of Platt Fields.

During early stages of data collection within the project, which took place in the project's first ecosystem case study of Manchester and Salford, parks quickly emerged as rich sites for participation, meaning and value. They were depicted as safe spaces of contemplation and access to nature, within interview narratives and ethnographic observation of people going to the park for restful solitude, for exercise or for other cultural practices, including singing, craft-making and feeding the birds and squirrels. They were spaces of sociability for connecting with family and friends, for team sports, recreation, and exercise, and for making new acquaintances. We heard powerful invocations of parks as places of nostalgia, where memories are formed and as places which conferred value onto their neighbourhoods. There were also narratives that articulated the rules and conventions for participation that varied between parks and within zones and areas of parks, between the urban, peri-urban and rural environments for the fieldwork and between social groups. The practices and behaviours of participation formed the social space of parks and informed their definition and their boundaries; some practices actively excluded others from parks and green spaces; some participants self-excluded themselves from specific areas or avoided parks at key times.

The target for the Understanding Everyday Participation (henceforth UEP) research was cultural policy, or rather the particular type of cultural policy within the United Kingdom from around the turn of the twenty-first century recognised by McGuigan, and diagnosed by subsequent cultural policy scholars as commodified (Gray 2007), defensively instrumental (Belfiore 2012), technocratic and conservative (Bunting et al. 2019) and politically opportunistic and ultimately divisive (O'Brien 2013; Brook et al. 2020). The political ontologies of cultural policy, we argued, displayed a dominance of economic rationalism, made possible through the performance management mechanics of New Public Management mechanisms. Public spending on arts and culture is primarily justified in return for non-cultural goals (Gray 2007) whilst relying on audience development models that appear to reinforce narrow, elite forms of arts and culture that neither attract nor satisfy most people in their everyday lives. This portrayal of the funded arts as a minority interest might seem contrary to the narratives of arts funding agencies and public bodies, who are at pains to state their purpose in creating great art for (and sometimes with) everyone (Arts Council England 2013). However, the evidence of participation data has repeatedly shown that there is a lack of engagement from large sections of the general population (Bunting et al. 2008), that those most prolifically involved in funded art forms are those from the most educated and affluent backgrounds, and that the large majority of others are likely too busy with other cultural and leisure activities to be able to take part, even if they wanted to (Taylor 2016).

In times of increasing social division, growing concerns about societal wellbeing; the challenges of polarisation, populism and cultural wars; and pressure on public funding, it is increasingly urgent to understand whose culture gets supported and which spaces are valued, maintained and held open in public trust. If only around 8.7% of the population (Taylor 2016, p. 169) regularly engage in state-endorsed and publicly funded forms of culture, then we should be reappraising what public policies for arts and culture are for. Definitions of cultural value matter because they are powerful. Naming something as 'cultural', or not, marks out boundaries of status and allocates or withholds resources for people and places. To identify what is valuable in people's everyday lives and to understand it culturally tells us not just how shared values are articulated and mediated but how prevailing structures of governance and lived experience of place are potentially affected, resisted and transformed through participation in cultural democracy and the practising of public citizenship.

This book investigates the connections and meanings that everyday participation in municipal parks has for cultural policy, place-making and governance, and for the practising and stewardship of public space. By adopting a critical cultural policy lens, it identifies a paradox for the municipal public park: as extraordinarily treasured places for active participation, attracting diverse communities and fulfilling a range of policy attachments and non-cultural policy objectives (Gray 2007), whilst held in public trust by municipal authorities, they have become increasingly difficult to maintain as accessible public spaces. Following the lead of the UEP project, the book aims to push the boundaries of 'culture', exposing the role these boundaries play in the making of economic, social and geographical inequalities, and proposing a progressive, more socially just, approach to cultural policy.

Chapter 2 introduces histories of parkmaking to consider how the values and policy rationales of municipal public parks have altered since their inception in Victorian England. It explores the political economy of parkmaking in the time of social upheaval, urbanisation and industrialisation, and explores its connections to class formation, social form and the cultural management of urban populations. Chapter 3 explores the parallel histories of museum-making and parkmaking, returning to John Ruskin and his followers whose art reforming advocacy bring the two together in Northern England in the nineteenth century. The practices of contemporary museums co-located in parks, and the thresholds between their different publics, are considered through two Manchester case studies. Chapter 4 considers the contemporary lived experience of the public park, presenting empirical research and discussion of the social lives of public parks and their users, drawing on analysis of qualitative data from household interviews, ethnographic fieldwork, stakeholder consultation and participatory workshops which took place in the six case study ecosystem sites of the UEP project. Chapter 5 considers the management of and advocacy for public parks as local cultural policy in the twenty-first century, identifying the 'public park paradox' which is at the heart of tensions between their principal owners, local authorities and retraction of funding and duty of care by the nation-state. It explores the lived experience and structures of feeling in the case study ecosystems, or their *cité* and their *ville*, examining the perceptions and participation of their residents in land management and stewardship. The concluding Chap. 6 returns to consider the right to the park as the right to the city, and the current context for the public park, as one of state retraction, commodification and depletion, threatening these treasured cultural public spaces of both commoning, and undercommons, through everyday participation.

Notes

1. This term circulated during the summer lockdowns of 2020 to describe the collective making of an outdoor space resembling a bar or public house, as venues remained closed but permitted numbers for gathering increased. My thanks to Dr. Susan Oman for bringing it to my attention.
2. Despite enquiries to the Ruskin Archive at the University of Lancaster, and extensive research on collections of Ruskin's writing, I have not been able to locate the primary source for this quote. It is likely to be a misattribution that has become an urban myth propogated by many park advocacy documents.

3. The term was reportedly first used as a descriptor for parks in London in the eighteenth century by William Pitt the Elder, Lord Chatham and two times prime minister, who was cited in Hansard by William Windham as he opposed a proposal in parliament to allow the crown to build on Hyde Park in 1808 (Crompton 2017, p. 111).

4. The project was funded by the Arts and Humanities Research Council Connected Communities 'Communities, Culture and Creative Economies' programme (AH/J005401/1) with partnership funding from Creative Scotland. For further information, see www.everydayparticipation.org.

5. These indicators were based on the national statistics on arts and cultural participation and review of actual and planned investment into arts and culture from local authorities, national lottery distributors and central government, as indicated by local budgets, numbers of regularly funded organisations and specific local policy initiatives. These formed the basis for selecting sites in England; two further case study sites were also volunteered by co-funder Creative Scotland. The case study sites were: Manchester/Salford; Newcastle-Gateshead; Dartmouth; Peterborough, Peterculter, Aberdeenshire, and the Western Isles of Lewis and Harris.

References

Ahearne, J., and O. Bennett. 2009. Implicit cultural policies. *International Journal of Cultural Policy* 15 (2): 139–244.

Arts Council England. 2013. *Great art and culture for everyone: A strategic framework 2010–2020*. 2nd ed. London: Arts Council England.

Barber, A. 1992. Letter: Public parks and spaces as a measure of a civilisation's greatness. *The Independent*, July 28. https://www.independent.co.uk/voices/letter-public-parks-and-spaces-as-a-measure-of-a-civilisation-s-greatness-1536236.html. Accessed 10 May 2023.

Barker, A. 2017. Mediated conviviality and the urban social order. *British Journal of Criminology* 57: 848–866.

Barker, A., A. Crawford, N. Booth, and D. Churchill. 2019. Everyday encounters with difference in urban parks: Forging 'openness to otherness' in segmenting cities. *International Journal of Law in Context* 15: 495–514. https://doi.org/10.1017/S1744552319000387.

Belfiore, E. 2012. "Defensive instrumentalism" and the legacy of new labour's cultural policies. *Cultural Trends* 21 (2): 103–111.

Bollier, D., and S. Helfrich. 2015. *Patterns of commoning*. Massachusetts: Levellers Press.

Brook, O., D. O'Brien, and M. Taylor. 2020. *Culture is bad for you*. Manchester: Manchester University Press.

Bunting, C., T.W. Chan, J. Goldthorpe, E. Keaney, and A. Oskala. 2008. *From indifference to enthusiasm: Patterns of arts attendance in England*. London: Arts Council England.

Bunting, C., A. Gilmore, and A. Miles. 2019. Calling participation to account: Taking part in the politics of method. In *Histories of cultural participation, values and governance. New directions in cultural policy research*, ed. E. Belfiore and L. Gibson, 183–210. London: Palgrave Macmillan.

Cohen, D.A., B. Han, K.P. Derose, S. Williamson, T. Marsh, L. Raaen, and T.L. McKenzie. 2016. The paradox of parks in low-income areas: Park use and perceived threats. *Environment and Behavior* 48 (1): 230–245. https://doi.org/10.1177/0013916515614366.

Colton, R. 2016. From gutters to greensward: Constructing healthy childhood in the late-Victorian and Edwardian public park. PhD Thesis, University of Manchester.

Conway, H. 1996. *Public parks*. Buckinghamshire: Shire Publications.

Cook, E.T., and A. Wedderburn. 1907. *The works of John Ruskin*, Library edition. London: George Allen.

Crompton, J.L. 2017. Evolution of the "parks as lungs" metaphor: Is it still relevant? *World Leisure Journal* 59 (2): 105–123. https://doi-org.manchester.idm.oclc.org/10.1080/16078055.2016.1211171.

Di Masso, A. 2012. Grounding citizenship: Toward a political psychology of public space. *Political Psychology* 33: 123–143. https://doi.org/10.1111/j.1467-9221.2011.00866.x.

———. 2015. Micropolitics of public space: On the contested limits of citizenship as a locational practice. *Journal of Social and Political Psychology* 3 (2): 63–83. https://doi.org/10.5964/jspp.v3i2.322.

Dobson, J., and N. Dempsey, eds. 2020. *Naturally challenged: Contested perceptions and practices in urban green spaces*. Springer Nature.

Elborough, T. 2016. *A walk in the park: The life and times of a people's institution*. London: Jonathan Cape.

Gehl, J. 2011. *Life between buildings: The uses of public space*. Washington, D.C.: Island Press.

Gidwani, V., and A. Baviskar. 2011. Urban commons. *Economic and Political Weekly* 46 (50): 42–43.

Gilmore, A. 2017. The value of parks and their communities: Research briefing. http://www.everydayparticipation.org/wp-content/uploads/2017/03/The-Value-of-Public-Parks-and-their-Communities-UEP-Research-Briefing.pdf. Accessed 20 Apr 2023.

Gilmore, A., and P. Doyle. 2019. Histories of public parks in Manchester and Salford and their role in cultural policies for everyday participation. In *Histories of cultural participation, values and governance. New directions in cultural policy research*, ed. E. Belfiore and L. Gibson, 129–152. London: Palgrave Macmillan.

Gilmore, A., and L. Lang. 2020. Talking, walking and making in Cheetham Park. *Conjunctions* 7 (2): 1–20. https://doi.org/10.7146/tjcp.v7i2.119258. ISSN 2246-3755.

Graham, H. 2023. Scaling heritage: Situated policy in an expanded ontology. In *Cultural policy is local*, ed. V. Durrer, A. Gilmore, L. Jancovich, and D. Stevenson. London: Palgrave Macmillan.

Gray, C. 2007. Commodification and instrumentality in cultural policy. *International Journal of Cultural Policy*. 13 (2): 203–215.

Greenhalgh, L., and K. Worpole. 1995. *Park life: Urban parks and social renewal*. London: Demos.

Habermas, J. 1992. *The structural transformation of the public sphere*. Cambridge: Polity Press.

Hardin, G. 1968. The tragedy of the commons: The population problem has no technical solution; it requires a fundamental extension in morality. *Science* 162 (3859): 1243–1248.

Harney, S., and F. Moten. 2013. *The undercommons: Fugitive planning & black study*. London: Minor Compositions.

Hewison, R. 2018. *Ruskin and his contemporaries*. London: Pallas Athene.

Hickman, C. 2013. "To brighten the aspect of our streets and increase the health and enjoyment of our city": The National Health Society and urban green space in late-nineteenth century London. *Landscape and Urban Planning* 118: 112–119.

Hodkinson, S. 2012. The new urban enclosures. *City* 16 (5): 500–518.

House of Commons. 1999. Memoranda submitted to the Environment Sub-committee of the Environment, Transport and Regional Affairs Committee on town and country parks, Session 1998–1999. https://publications.parliament.uk/pa/cm199899/cmselect/cmenvtra/477/477mem18.htm. Accessed 25 Apr 2023.

Jones, K.R. 2018. 'The lungs of the city': Green space, public health and bodily metaphor in the landscape of urban park history. *Environment and History* 24 (2018): 39–58. https://doi.org/10.3197/096734018X15137949591837.

Jones, K.R., and J. Wills. 2005. *The invention of the park: From the garden of Eden to Disney's Magic Kingdom*. Cambridge: Polity Press.

Layton-Jones, K. 2018. Manufactured landscapes: Victorian public parks and the industrial imagination. In *Gardens and green spaces in the West Midlands since 1700. West Midlands Publications*, ed. Malcolm Dick and Elaine Mitchell, 120–137. Hertfordshire: University of Hertfordshire Press.

Lefebvre, H. 1968. *Le Droit à la ville*. Paris: Anthropos.

Linebaugh, P. 2014. *Stop, thief!: The commons, enclosures, and resistance*. California: PM Press.

Low, S., D. Taplan, and S. Scheld. 2009. *Rethinking urban parks: Public space and cultural diversity*. Austin: University of Texas Press.

Mackay, D. 2010. New commons for old: Inspiring new cultural traditions. In *End of tradition? Part 2 commons: Current management and problems (cultural severance and commons present), Landscape archaeology and ecology*, ed. I. Rotherham, M. Agnoletti, and C. Handley, vol. 8, 109–118.

Manchester City Council. 2021. *State of the city report 2021*. Manchester: Manchester City Council.

Massey, D. 2005. *For space*. London: SAGE.

McGuigan, J. 1992. *Cultural populism*. London: Routledge.

———. 2004. *Rethinking cultural policy*. Maidenhead: Open University Press.

———. 2009. *Cool capitalism*. London: Pluto Press.

Mell, I. 2022. Examining the role of green infrastructure as an advocate for regeneration. *Frontiers in Sustainable Cities* 4: 1–21. https://doi.org/10.3389/frsc.2022.731975.

Miller, T., and G. Yudice. 2002. *Cultural policy*. London: SAGE.

Mouffe, C. 2005. In *For an agonistic public sphere in radical democracy: Politics between abundance and lack*, ed. L. Tønder and L. Thomassen, 123–132. Manchester: Manchester University Press.

———. 2013. *Agonistics: Thinking the world politically*. London/New York: Verso.

National Heritage Memorial Fund, and Heritage Lottery Fund. 1996. *Annual report and accounts of the Heritage Lottery Fund and the National Heritage Memorial Fund 1995–1996*. London: National Heritage Memorial Fund.

Nutter, R. 2020. *Paradise is here: Building community around things that matter*. Sheffield: Guild of St George.

O'Brien, D. 2013. *Cultural policy: Management, value and modernity in the creative industries*. London: Routledge.

O'Reilly, C. 2019. *The greening of the city: Urban parks and public leisure, 1840–1939*. New York: Routledge.

Ostrom, E. 1990. *Governing the commons: The evolution of institutions for collective action*. Cambridge: Cambridge University Press.

Pykkonen, M., N. Simanainen, and S. Sokka, eds. 2009. *What about cultural policy?* Helsinki/Jyvaskyla: Minerva.

RARE. 2020. *RARE newsletter*. Vol 23, Iconic Indian Cities, December 2020, Rare. https://rareindia.com/newsletter/37. Accessed 25 Apr 2023.

Rifkin, J. 2000. *The age of access: How the shift from ownership to access is transforming modern life*. London: Penguin.

Ruff, A.R. 2016. *Manchester's Philips Park: A park for the people, by the people, since 1845*. Stroud: Amberley Publishing.

Sennett, R. 2018. *Building and dwelling: Ethics for the city*. London: Allen Lane.

Shoari, N., M. Ezzati, J. Baumgartner, D. Malacarne, and D. Fecht. 2020. Accessibility and allocation of public parks and gardens in England and Wales: A COVID-19 social distancing perspective. *PLoS One* 15 (10): e0241102. https://doi.org/10.1371/journal.pone.0241102.

Smith, A. 2013. *Events in the city: Using public spaces as event venues.* London: Routledge.

Standing, G. 2019. *Plunder of the commons: A manifesto for sharing public wealth.* London: Penguin.

Tandy, V. 2019. The heritage lottery fund and its role in the construction and preservation of the past: 1994–2016. PhD thesis, University of Manchester.

Taylor, M. 2016. Nonparticipation or different styles of participation? Alternative interpretations from taking part. *Cultural Trends* 25 (3): 169–181.

Ward, D. 2014. *The commons in history: Culture, conflict, and ecology.* Cambridge: The MIT Press.

Wellington City Council. 2009. Wellington City Council public space design policy December 2009. https://wellington.govt.nz/-/media/your-council/plans-policies-and-bylaws/plans-and-policies/. Accessed 25 Apr 2023.

Williams, R. 1984. State culture and beyond. In *Culture and the state,* ed. Lisa Appignanesi, 3–5. London: Institute of Contemporary Arts.

———. 1988. *Keywords: A vocabulary of culture and society.* London: Fontana.

CHAPTER 2

Parkmaking, Municipalisation and Cultural Policy

Abstract This chapter examines the origins of municipal public parks in England and the policy rationales and actors that lay behind their establishment. It considers the symbiosis between the political economy of parkmaking and broader strategies of cultural management, the development of civic institutions and the management of the city as a public body. It highlights the absence of everyday park users within the historiography of public parks and sets out the policy discourses which informed the founding and funding of parks and their emergent business models, and which represented the political and class interests of parkmakers, designers and advocates. The chapter finds that along with a number of civic and cultural institutions, the making of public parks in the nineteenth century was not only an important strategy for social and moral improvement, managing public space and urban populations but also a fundamental form of municipalisation that saw the responsibility of the public park placed firmly in the hands of local government. The chapter also identifies a paradox whereby the legacy of Victorian parkmaking and its political economy has undermined the sustainability of park business models despite the strong case for public ownership and management in contemporary policy.

© The Author(s), under exclusive license to Springer Nature
Switzerland AG 2023
A. Gilmore, *Culture, Participation and Policy in the Municipal
Public Park*, Palgrave Studies in Cultural Participation,
https://doi.org/10.1007/978-3-031-44277-3_2

Keywords   History of parkmaking • Civic institutions • Cultural institutions • Municipal government • Philanthropy • 'Green lungs' • Policy rationale • Rational recreation • Social reform • Political economy

INTRODUCTION

Public parks came into being at a time of great social change within cities and towns in the UK. The twin midwives of industrial revolution and public reform prompted a remarkable period of innovation and expansion of many urban civic institutions we now take for granted. Historians and scholars of the city have illuminated how the toxic shock of these societal changes, driven by technology but powered by evolving forms of capitalism, forced policy responses to counter and control the unexpected or unintended consequences of progress. As Peter Hall describes as an historical understanding of urban planning:

> Time and change happeneth – within limits – to us all. But the limits are real: finally it is the technological-economic motor that drives the socio-economic system and, through it, the responses of the political safety-valve. (Hall 1996, p. 4)

Reflecting on the birth of the public park in hindsight reveals its contingency with other contemporaneous changes in the urban realm and governance of the newly industrialised city. The mobility and movement of populations, the expansion of manufacturing and distribution, hard infrastructure of transport, civic buildings and housing and the concomitant creep of new neighbourhoods out into the surrounding countryside set the scene for class formation and socio-cultural distinction. The industrial cities produced public cultures of display and spectacle specific to the urban identities that were in formation, distinguishing class, status and politics and articulating power and agency (and their lack) for the new middle classes and urbanised working classes (Gunn 2007). Their histories perceive the constant threat of political unrest, disease and poverty, crime and the newly assembled mob of the working classes contesting space and requiring moral regulation in order to maintain economic stimulus and stability. The philosophies for moral improvement proposed by poet, educationalist and critic Matthew Arnold in *Culture and Anarchy* ([1869]

1981), which have been so influential on cultural policies and theories of democracy (Bennett 2005), were motivated by concerns of disorder visibly illustrated by the Hyde Park riots of 1866. The threat of anarchy, for Arnold, needed immediate counterbalance through cultural education and the management of the public body, a part in which municipal public parks were destined to play.

The establishment of parks for public use in these 'shock' industrialised cities (Briggs 1963) responded to middle-class anxieties about the impact of the urban condition on the morality and health of the working population (Sigsworth and Worboys 1994; Wyborn 1994). During the nineteenth century, the accelerated process of urbanisation brought about by the industrial revolution produced a profound demographic and environmental impact on British towns and cities (Barker 2004). This was particularly the case in the north-west of England, where the heartland of the industrial revolution, Manchester, experienced a rapid population expansion from 75,000 people in 1800 to 2,117,000 in 1900 (Jerram 2011, p. 2). The development of urban centres for manufacture and for housing the working populations led to the movement of wealthier classes to the suburbs, enclosing former agricultural land. As the countryside became more distant, an interest in providing access to open spaces was transplanted into the city, promoting the right to roam and escape from city squalor in response to urban ills. It instigated the lobby for the provision of grounds for walking which were within reach and accessible to the working population, and in February 1833, a House of Commons Select Committee considered how to improve access to open space in order to promote health and comfort (HCPP 1833, p. 2). The committee's report informed national strategy targeting the new and expanding manufacturing centres by providing public walks and open places and establishing the moral case for parks, based on a series of economic and public health arguments (Gilmore and Doyle 2019).

This chapter examines the policy rationales given historically for the establishment of urban parks under municipal authority through the lens of cultural policy, as established in Chap. 1. It explores these rationales within the context of the political economy of park design and establishment and the values and principles associated with the models for their operation and management. It builds on and extends the discussion of public park historians, such as Conway (1996, 2000), Ruff (2016) and O'Reilly (2013, 2019), of the motivations and concerns for parkmaking and administration and their connections to urban politics and policy, drawing on secondary sources, archival research and case studies of parks in the North West of England, to identify the values attributed to parks by their advocates and engineers.

In the first section, I examine how parks offered a strategy for public health promotion closely aligned with contemporary medical practices and beliefs, through the metaphor of the social body of the city (O'Reilly 2019) and the role that parks play as its lungs in providing oxygen to individuals and community. The importance of providing open recreation space for a healthy workforce is the most powerful and durable rationale for public parks, one commonly repeated today. However, the origins of this rationale lie not in contemporary medical evidence but in the Romantic Movement of the late eighteenth century, which in the early nineteenth century became a feature of urban rather than rural concern (Howkins 2011). Furthermore, the metaphor proposed moral as well as physical health, both in the sense of propriety and civic responsibility to the public, on the part of parkmakers, and in their aspirations to inspire proper healthy pursuits amongst the urban working classes.

Exploring the ritual practices and performances documented within historical accounts of the Victorian public realm, I consider how parks were spaces that both contained and permitted particular formations of class and social distinction. The establishment of public parks provided the means through which to instigate and mediate cultural participation to govern the cultural practices that took place within urban public space. The chapter introduces how the aesthetic logics of park design were intertwined with the economic motivations and elite interests of the time. The history of parkmaking is one that reveals the governmental processes of "bounding, design and regulation" (Booth et al. 2020, p. 3) to make "spaces apart" (Booth et al. 2020, p. 3), and the chapter examines how parks demarcated urban space and established boundaries afforded by their design and managed through their regulation.

The chapter considers the political economy of parkmaking, from the motivations and capitals of those involved in advocating, planning and managing public parks. It traces the legacy of these parkmakers on the funding models, design thinking and policy transfer for municipal authorities, from the first public urban parks in the nineteenth century to the present day. Undertaking a historiography of parkmaking reveals tensions between the intentions of policy-makers to create 'parks for the people' and the sustainability of keeping parks public, which is jeopardised by their inherited business models. It also favours the interests and aspirations of elite individuals and social groups, obscuring and marginalising other voices and histories. As O'Reilly (2019) notes, a key element missing from the histories of public parks is the voice of the park user. The archival bias means historical accounts predominantly focus on those involved in lobbying for and designing parks, and to a lesser extent those who maintain parks, rather than their everyday users. Where the user is present, a classed dimension is found within the archive in the contribution of individual agitators and advocates who made their mark on the historical record, for example, through editorials, letters and articles in the popular press, design competitions, administrative documents, publicity pamphlets and committee minutes. These were people who had the cultural and social capital to be part of the cultural public sphere of parkmaking: everyday users are only recognisable *en masse* through their participation in events, protests or other recorded assemblies or as impersonalised visitor figures. As individuals, they are primarily only visible as criminals, identified within court records, park keepers' journals and newspaper articles on park misdemeanours.

Dreher (1993) views this absence as one that undermines the contingent relationships between parks, their histories and the continual making of the cities in which they are based, finding that the majority of British park histories neglect "how public parks became integral and positive elements of the city and its culture" (Dreher 1993, p. 9). Her research on the later period of parkmaking between 1870 and 1920 shows how studying participation and user management in public parks reveals their democratic potential and the practices of social display, which shaped public culture through their central role in city living. Here, I also mark the absence of the everyday user and of working-class voices within the histories of parkmaking. I also support Dreher's concerns to refocus on the relations between sites of public parks, their place contexts, and the social relations surrounding their production, tracing connections between the policy rationales provided for their provision within the cultural ecosystem of the city, and the gestures, discourses and practising of public space articulated through everyday participation.

The Origins of the Public Park

As Elborough remarks in an engaging narrative history of parks, the public park is a recent invention with a much longer provenance in land designated for the private use of raising animals for hunting (Elborough 2016).

He traces the earliest description of what we might now call a park to 2000 BC, when hunting preserves were cultivated to create luxurious habitats for beasts and huntsmen in ancient Assyria. The deer park was also part of Romano-Britain and became instituted within mediaeval England post-Norman conquest through Forest Law in 1079 as the organisation of land ownership, and function for mediaeval manors brought new regulatory and administrative processes. Emparkment was essentially the enclosure of land for the purposes of hunting under royal licence, designating rights to the private few given licence and establishing severe penalties, including death for deer poaching, for those who broke the Forest Law. There were 25 royal forests recorded in the Domesday book, rising to 143 in 1215, As more land became enclosed, its ownership and use rights were distributed to the Norman aristocracies and customary common rights to forage, graze and make other uses were outlawed (Standing 2019, p. 9). These common rights were however reinstated by the Charter of the Forest in 1217, which established the basis for access to 'disafforested' areas, along with new roles for stewardship of the land and its beasts. Charters set out what were effectively rights to subsistence for commoners, through detailed descriptions of the activities that could take place within agreed boundaries, such as gathering firewood, cutting peat, feeding pigs with acorns and fishing, along with the fees due to the lord of the manor. These common rights persisted for hundreds of years, and the Charter became the foundation for common law and civil rights beyond the advances made by the better-known Magna Carta (Standing 2019).

Emparkment and the subsequent social contract by charter created processes through which the practices of the non-elite were policed and spatially bound. They also produced ritualised behaviours as hunting practice became more managed, observable and competitive. These parks became arenas of display and etiquette, acquiring landscaping and built structures, and founding new farming practices, land management skills, and sports that transferred into other areas of everyday life.[1] Throughout the mediaeval period, parks continued as an adjunct to the manorial estate, becoming more elaborate and cultivated: "the private park was emerging as a much broader aristocratic theatre of play, one where hunting in all its forms would slowly start to take a back seat to horticulture" (Elborough 2016, p. 21).

The post-mediaeval centuries saw an expanded function of these private parks. Previously hidden behind formal gardens, they became conjoined and landscaped in order to present the estate owner or visitor with an extended view of their status and power. Like the mediaeval deer park, they simultaneously reproduced and obscured social and economic divisions, and would cede the aesthetic inventions and horticultural techniques that would, in turn, become an important influence on municipal public parkmaking. The architects of these parks, such as Capability Brown and Humphrey Repton, presented the aristocracy with a managed purview of their wealth, romantic vistas that separated their estates from the city as "secluded, private and rural" (O'Reilly 2019, p. 2). From the seventeenth century, some gardens and royal parks were available for public uses, such as walking, sharing space with the elite pursuits of racing and hunting (O'Reilly 2019), although successive monarchs gave orders to enclose part or all their land to keep out commoners (Conway 1996). Parks such as Hyde Park and St. James Park became the stages for further display of aristocratic taste and fortune, not least in the horticultural fashions they spawned which influenced the design of private gardens (Elborough 2016). They were supplemented by new 'pleasure grounds' that offered outdoor entertainment on a commercial basis, effectively keeping out those who could not afford entrance but also sustaining revenue models through admissions charges, commissioned entertainments and sales of refreshments, which also formed a route into the park for the lower classes who could not afford entry. Grounds such as New Spring Gardens and Vauxhall Gardens were not just semi-public spaces but sites for the performance of cultural practices and social mores, which crisscrossed the boundaries of private and public, regulatory and entrepreneurial, encompassing masked balls, balloon ascents, side shows and circus acts alongside more mundane practices of promenade, prostitution, drunken revelry and petty crime (Elborough 2016).

The history of parkmaking runs parallel to the expansion of private capital and the infrastructures of public management. It was only as land enclosure became part of national industrial strategy that the universal right to access parks became a policy issue. The effects of the enclosure movement are some of the most hotly debated in English social history (Howkins 2011), with claims that enclosing the open field system of agriculture fundamentally changed class structure, destroying the peasant class and forcing wage labour, contrasting with those that the commons subsidised an existing working class. The majority of land enclosures took place prior to 1830, but by the mid-nineteenth century, enclosure was predominantly associated with the industrial urban context. The enclosure of the commons became an enduring concept for radicalisation and agency, symbolising an English poor, robbed of their birthright. 'Commoners' who used the land for grazing or firewood collection were marginalised, and the commons became less a productive resource and more public amenity, providing freedom to roam and open space, allied to the concerns over a lack of provision for public walks. New lobbying groups established to protect the commons incorporated the cause into social policy agendas: "enclosure was a moral issue – it represented the conspicuous consumption of land stolen from the people by a useless aristocracy" (Howkins 2011, p. 127).

The urbanisation of the industrial revolution diminished open space for recreation inside cities. A radical movement concerned with preserving the commons joined a new urban interest in access to green space to escape the squalor of the city, which had roots in the Romantic Movement of the late eighteenth century (Gilmore and Doyle 2019). This, in turn, spawned advocates for public subscription and rate-raising for accessible open green space and private enterprise which hoped to profit from the value of parkland to those who could afford it. As the wealthy moved out to the suburbs, the entrepreneurs moved in, building housing developments that attracted them through their proximate green spaces, such as Regent's Park, London, Prince's Park in Liverpool and Victoria Park in Rusholme, Manchester. These parks were financed through the sales of these high-end housing estates, whose owners had privileged access. Whilst they influenced new architectural styles and created pockets of retained green space, their financial model ultimately failed with land passing from the private trusts to public authorities but usually with no funds for ongoing maintenance and sites that were in poor condition (Layton-Jones 2016).

The transfer of private estates and costs to public hands is a universal story in parkmaking. In Manchester, three public parks were originally large working estates and private parkland owned by the aristocracy in the eighteenth and nineteenth centuries: Platt Fields and Wythenshawe Park in South Manchester and Heaton Park to the North. The grounds for mansion houses and other estate buildings with their own subsequent functions, these parks have much longer histories connected to landscape management, ownership and use, which navigate the boundaries between commoner and elite, public and private. Wythenshawe was the site of an early mediaeval deer park as part of the Cheshire Tatton demesne, whilst Platt Fields is transected by a tenth-century defensive ditch and boundary marker. All three were recipients of remodelling that connected their stately houses with green spaces for private pleasure and presented romantic extensive vistas of their domains, whilst hiding the agricultural labour that powered their estates. Both Heaton and Platt benefited from the landscaping of William Emes, who brought pleasure-ground designs and naturalised horticultural effects, such as ha-has and tree clumps, influenced by Capability Brown (Latimer 1987).[2] Wythenshawe's gardens were re-designed in the 1850s by John Shaw to frame the existing landscaping to the front and rear of the Hall (Historic England 1986). The land did not become a public park until 1926, when it was purchased and donated by local industrialists and public servants, Ernest and Shena Simon (later Lord and Lady Simon of Wythenshawe), an act which played a critical part in the development of Wythenshawe into a suburban 'Garden City' social housing estate in the 1930s (Olechnowicz 2000). In Manchester as in London, parkmaking was linked to private housing developments: in the 1830s, as wealth grew in the city, companies were formed to share interest in available land and new parks, such as Victoria Park, became part of their business model. These new estates were managed through private associations and were effectively gated communities: Victoria Park is described as "[in 1845] a private, self-governing rural retreat for merchants and some professional people living in Manchester" (Spiers 1976, p. 20).

The same decades of the 1830s and 1840s saw the growth of commercial pleasure grounds in Manchester, such as the Manchester Zoological Gardens in Broughton, Belle Vue in Gorton and the later Pomona Gardens, which remained operating until the turn of the century (Latimer 1987). Following the earlier Tinker's Gardens in Salford, these featured zoos and menageries, performances, panoramas and other exotic delights, alongside outdoor pursuits such as archery, hare coursing, and bowling, and were attractive as green spaces in the city that offered all the "pleasures of a rural fete without the expense of a railway trip" (Chetham's Library 2013). These grounds had relatively long-standing commercial success, unrivalled by other activities (except each other) through the range of amusements on site and their accessibility to those in the city. Whilst their entry costs presented a barrier to some, they provided hugely popular leisure spaces that saw increasing social mixing of classes until their decline in the late nineteenth and early twentieth centuries (Powell and Wyke 2015).

The promotion of urban green space by public policy was not driven by an interest in individual pleasures or intrinsic experiences but by instrumental concerns, which in contemporary cultural economics would be understood as positive externalities (Bennett 2005) and policy attachments (Gray 2002). The 1833 House of Commons Select Committee on Public Walks shaped national strategy for tackling the case of the sprawling manufacturing centres by the provision of open space that "would much conduce to the comfort, health and content of the classes in question... [and] Public walks may be gradually established in the neighbourhood of every populous Town in the Kingdom" (HCPP 1833, p. 3). This report established the moral case for parks, gathering a series of economic and public health arguments in favour of the creation of freely accessible open urban spaces, a case dominated by the discourses of public health that were closely allied to broader philosophies of the city (O'Reilly 2019). City life at this time entailed the polluting influence of its own population, the health of which was in turn shaped by its surrounding urban environment. Joyce (2003) characterises this as a necessary equilibrium required between the "sanitary economy" (p. 65) of the town and that of the self: the social body of the city, like the human body, was made of particles that circulate through different environments and organs. This metaphor was established in the popular imagination by the naming of parks as 'the lungs of the city', a description attributed in a parliamentary speech by a Mr. Windham, to William Pitt the Elder, who called parks "the lungs of London" when discussing the threats of development to Hyde Park in 1808 (O'Reilly 2017, p. 5). The biotic reference mirrors other bodily references to the urban infrastructure and the remedy of its ills through the application of technologies and strategies for sanitary reform and spaces for recreation, in what Jones calls "somatic urbanism" (Jones 2022, p. 1207). As trees and greenery serve to oxygenate the environment, it was hoped parks would provide the public realm to refresh the physical, moral and psychosocial health of the 'humbled classes' (House of Commons Parliamentary Papers 1833). Parks were a gift to the poor, offering sanitary opportunities for outdoor exercise that could substitute and reform their otherwise immoral practices. As Dr. Kay, the Manchester physician and vocal element of the city's local parks lobby, stated in evidence cited in the Select Committee:

> At present the entire labouring population of Manchester is without any season of recreation, and is ignorant of all amusements, excepting that very small portion which frequents the theatre. Healthful exercise in the open air is seldom or never taken by the artisans of this town, and their health certainly suffers considerably from this deprivation. The reason of this state of

the people is, that all scenes of interest are remote from the town, and the walks which can be enjoyed by the poor are chiefly the turnpike roads, alternately dusty or muddy. Were parks provided, recreation would be taken with avidity, and one of the first results would be a better use of the Sunday, and a substitution of innocent amusements at all other times, for the debasing pleasures now in vogue. I need not inform you how sad is our labouring population here. The health of the lower classes is much depressed by the combined influence of municipal evils, and their own corrupted manners and constant toil; but the total absence of all honest sources of amusement, and the neglect even of healthful exercise, are features in which I would fain hope we are singular. (House of Commons Parliamentary Papers 1833, p. 66)

The practice of walking was part of a "cultural framework" (Solnit 2000, p. 83) popularised at the turn of the previous century by the poetry of Wordsworth, who also personally promoted the need for spaces for public walking for city dwellers (Crompton 2017). The 'green lungs' metaphor became an enduring metaphor, shored by its moral and classed associations with the romantic pleasure of walking in nature. It articulated a significant relationship between individual and wider community health that established the dual function of parks in serving individual private good, through access to outdoor recreation and generating public good by bringing to the city air the oxygen released by their vegetation (Crompton 2017). It was homologous to contemporary medical practice, and many early advocates for public parks came from the medical community. Epidemiology at the time made links between the acreage and availability of parks and green spaces to the urban poor and local death rates, particularly in the time of epidemics, in cities such as Liverpool, Manchester and Salford (O'Reilly 2019). There was a common acceptance of miasmas as the primary explanation for infection and disease[3] which buoyed the ease with which the lungs metaphor, and with it the urban parks movement, transferred overseas, to other city planning contexts such the US and Australia (Crompton 2017). However, the efficacy of parks as a remedy to polluted air in the industrial city met its match in the realities of heavily polluted Manchester, when it was questioned even as Queens and Phillips Parks were established in 1846 (Ruff 2000).

Even so, within two years, the 1848 Public Health Act provided voluntary empowerment to local authorities for the establishment of parks and other public realm works for health reasons, and Britain's towns and cities saw many new civic spaces from public squares to shopping arcades built from this time (O'Reilly 2019). The emphasis on healthiness as a moral attribute, demonstrated by Dr. Kay's witness statement, suffused the discourses and regulation of other areas of public life and leisure time. The creation of dedicated, managed spaces for public walking was a means for the absorption of working-class leisure time into healthy pursuits, drawn away from the hidden amoral spaces of the public house and private homes into visible 'rational recreation' (Wyborn 1994). Public health provided, therefore, the core policy rationale for public parks enshrined within national legislative frameworks and realised locally. The Public Health Act, followed by further acts in 1875 and 1925, carried not only the development of local sports and recreation facilities but also the instruments for social policy and moral proscription, accumulation of political capital and wealth, as discussed in the following sections.

PERFORMANCE AND RITUAL IN PUBLIC SPACE

Parks were one of a number of institutions that offered the opportunity for 'social engineering' in the mid-nineteenth century (O'Reilly 2019). Museums, art galleries, reading rooms and libraries formed the spaces for a civic public sphere that facilitated governmental strategies of discipline and surveillance, shaping bodily comportment via exhibition design, architectural devices and their effects on the public body (Bennett 1995). Gunn (2007) defines the period of the 1840s to 1880s as an exceptionally intense period for moral regulation, identity formation and performance in the public realm, enacted through dress, comportment and gesture in central city spaces. He discusses this as a social phenomenology which brought with it hierarchies and regulatory orders. Like the metaphor of green lungs, the idea of the moral city was both representational and regulatory of the public body. The city is "inseparable from its designation as a space of physical and moral danger" (Gunn 2007, pp. 60–61); morality, physical wellbeing and the functioning of the city as ecosystem become conjoined through the placing, movement and perception of bodies in the public realm, turning city recreational spaces into a "monumental stage set" (2007, p. 66) for ritualised performance.

Social order in mid-Victorian industrial cities was explicitly organised around mental labour and manual/menial labour: the brain of the middle classes and the brawn of the working classes. In the new public spaces of cities, such as parks, streets and squares, the ritual movements through city spaces signified class and status. These acts were mainly performed by bodies at leisure, as working hours and spaces kept the labouring body moral and contained with the threat of moral vacuum unleashed when the factory gates closed. The daily regular enactments of everyday life served to identify and classify social types, through the types of activity undertaken and the routes and spaces inhabited. For example, Gunn (2007) shows how within this period the middle class deserted the city centre, moving their residence to the suburbs or to villages such as Altrincham and Alderley outside of Manchester. The movement of commuting into and around the centre, the modes of transport taken and times and the specific days when in motion established patterns which indicated class and difference. Gunn identifies a weekly cycle of activities which included the "High Change" when business and trading took place at the Manchester Exchange on Tuesdays, amidst the daily rituals of promenade around the city centre shopping streets, galleries and libraries. These rituals favoured specific spaces: in Manchester 'Doing the Square' around St Annes and Kings Street allowed the recognition of others within and outside your social type (Gunn 2007, p. 75). Other ritualised promenades around set routes are documented in accounts of Manchester youth provided by Charles Russell (1905) who described the 'Monkey Parade' as a regular Sunday night occasion for working girls and boys to take part in acts of courtship. This involved gangs promenading in parallel single-sex groups around set routes, hanging out on corners, cross-street heckling and dress codes such as the wearing of button-hole flowers. The practice persisted through the second half of the nineteenth century until at least the 1930s (Haslam 1999) and has been reported elsewhere even more recently, for example in the nearby town of Macclesfield where it is known as the Monkey Run (Gilmore 2013).

Such practices must have frustrated the aspirations of public realm planners who wanted to reassure the social body that city spaces were free from moral and physical danger. The pace of urban development to house new workforces had resulted in unplanned neighbourhood slums and 'no-go' zones ringing the protected private property of city centre commercial districts, where criminal and amoral pursuits could be encountered. Whilst the dangers and sins of these streets may have been romanticised and even dramatised within local press accounts, they formed the symbolic social and spatial boundaries between the proletariat and bourgeoisie, further signalled by ritualised performance. Working-class participation demanded containment, therefore, and the strategies most called upon were derived specifically for public health and involved social surveillance through the mapping and designation of public space. The "moral mapping" of the city (Gunn 2007, p. 63) can be seen in various published cartographic forms, for example, the Manchester Drink Map of 1889 (see Fig. 2.1), which detailed the places where alcohol and its associated pursuits could be found in the city, the sites where the spatial and social boundaries between respectability and immorality may be breached.[4] Despite the rise in reporting and collation of crime statistics, policing was not fully trusted by the middle classes to look after private property or public space comprehensively until 1880s, and "means other than policing were necessary to bring a sense of social order and stability to the city streets" (Gunn 2007, p. 65).

PARKS AS POLICY SPACES

Public parks were therefore a remedy; presenting accessible spaces which allotted urban leisure time, albeit more compressed for the working classes, as one of a number of new civic institutions that contained the public body. The museum, reading room, library and art gallery were places that offered varying spatial logics of practice and forms of embodied expression. Municipal museums were established over the same period as public parks and shared similar policy rationales following lobbying and legislation led by 'middle-class radicals' in the 1830s and 1840s, influenced by political reform and the tensions of the period relating to political unrest, industrialisation, overcrowding and public health. They presented opportunities for walking, gathering and sitting, which alongside their social and educative missions were mechanisms through which to manage publics and produce the social body (Rees Leahy 2012). Hill's (2005) analysis of the development of municipal museums identifies a wider project of class formation and position (as is discussed further in Chap. 3). As Hill outlines, these cultural policies ostensibly targeted the improvement of the working classes, in common with parks in providing the means for middle-class legitimacy and authority, whilst also extending opportunities for leisure and taste cultures that could serve to exclude those of lower social status:

> [I]t is clear that initiatives intended for the benefit of the public as a whole could easily become exclusively middle-class: parks, Mechanics' Institute and exhibitions, to take some examples, were either situated in middle-class areas, or imposed *de facto* restriction through dress code or price, or offered events that the working class could not or would not appreciate. (Hill 2005, p. 40)

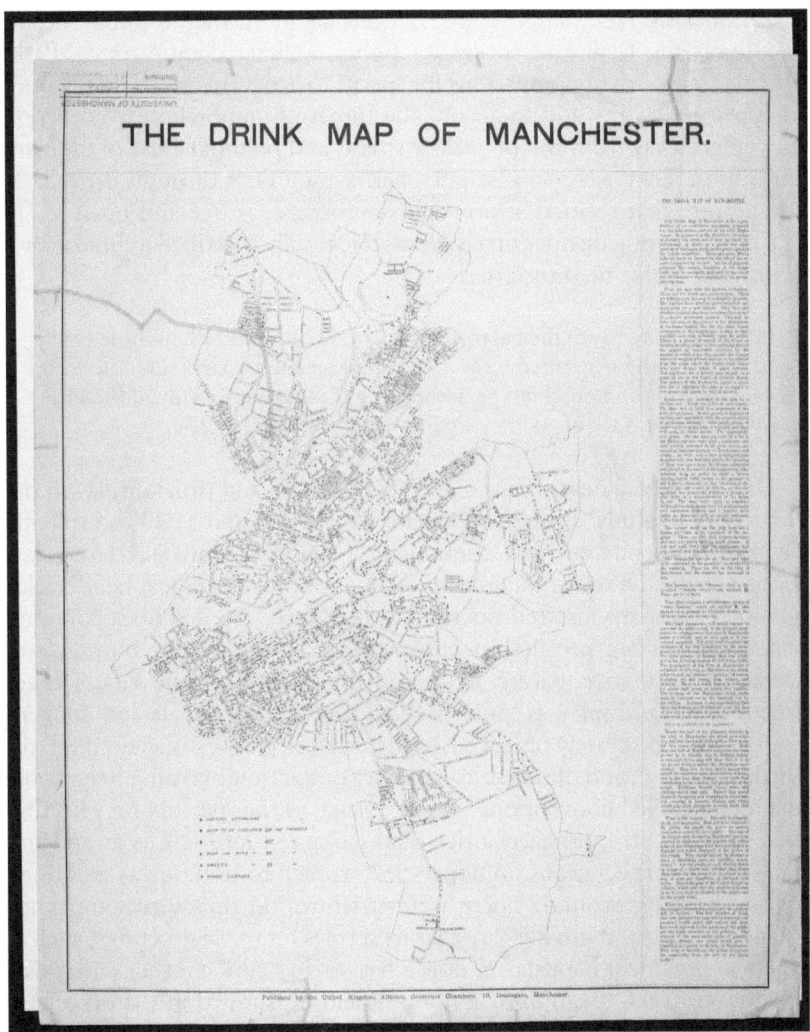

Fig. 2.1   The Manchester Drink Map. (Image courtesy John Rylands Research Library. © The University of Manchester)

Parkmaking has therefore always been far more than a public health service, as the facility of spaces for public walking demonstrates. Parks provided new urban populations for spaces to articulate social distinction, to construct civility and to absorb the threat of immorality by attracting park-goers away from the unhealthy streets and polluted areas of industry and slums. They were stagesets for performing class through dress code and gesture that seemed in principle democratic, as free and open to all, but also rendering distinction visible, as bucolically described in Bradshaw's *Illustrated Guide to Manchester*:

> It is interesting to see the happy tradesman and his wife, with his little family by his side, the contented weaver and the Manchester clerk and salesmen, with their entire household perambulating these grounds with all the hilarity and importance of country squires. (Bullock 1857, p. 22)

Dreher (1993) describes the practices of social and ritual display in the period of her study of public parks between 1870 and 1920 as explicitly classed and gendered, but contingent with interaction and cross-class social mixing. Drawing on sources such as etiquette guides, handbooks, memoirs and park-inspired poetry and literature, she describes common park uses that are private-but-public: lovers' trysts, family outings and access to restorative nature to escape the ravages of the city. Dreher describes social display as 'publicity': to be effective, rituals and conventions need to be closely observed, particularly for the aristocracy and 'fashionable Society', and observable in public. Etiquette-governing behaviours included detailed conventions for dress, such as the specifics for selecting men's 'park suits' and accessories (dark-coloured morning coat and top hat) and the expectations on upper-class women to don outfits fashioned for the hunting ground. There were strictures on the correct times for walking, riding and driving carriages, and rules for social greetings, such as when to provide a handshake, doff a hat, or to "cut" another park-goer, to show critical evaluation by ignoring them. Enshrined in visitor guides, handbooks and park history books, particularly for London Society, these public gestures demanded an audience, as well as a readership, and offered routes for social climbing and mobility if correctly adopted:

> Public parks offered the perfect theatre for the show of Society, allowing Society to define its boundaries and individual citizens to perform for a knowledgeable and critical audience. The growing importance of the public

community, and of public space, meant that while fashionable Society tried to create a special identity, attempting to establish a private zone within the broad park culture, its members nevertheless craved public attention and approbation. (Dreher 1993, pp. 164–165)

A walk (or ride) in a park could therefore demonstrate respectability and display social aspiration, as could the choices of recreation undertaken for more active users. Parks are not protected bubbles of public space but absorb and reflect the social innovations and ideological shifts around them. Victorian and Edwardian practices of park use responded to technological developments, such as the popularisation of cycling in the 1890s and the incursion of the motor car in the 1900s, both of which affected the dress code and time-zoning of social display, and also the gendered spatial integration and flow in the larger parks, such as Hyde Park, which could accommodate these different forms of traffic (Dreher 1993). New parks outside of the capital strove to facilitate similar behaviour: Queens Park, Philips Park and the later Alexandra Park all had carriage driveways laid out in their original plans, although there is little evidence that these were used beyond the more popular promenading (Latimer 1987) (Fig. 2.2).

The character and amenities within parks are subject to demand but also opportunity, shaped by the broader structures and nodes of political economy within the wider urban ecosystem. Conway (1999) outlines the expediency of 'second wave' inner city parkmaking developed following a trio of legislative acts that granted local authorities powers to transform urban land for public good. Municipal authorities were empowered by the Public Health Act of 1875, followed by the Open Spaces Act 1881 and the Disused Burial Grounds Act 1884. These acts enabled the authorities to turn disused burial grounds when overcrowded and in disrepair, following the Burial Act of 1853, into football pitches, sports grounds and playgrounds with more emphasis on amenities to support physical recreation rather than engagement with nature or education. The sites were flattened, headstones moved to one side, and rapidly converted with turf, playground equipment and other infrastructure to provide spaces that did not require expensive landscaping but which offered rational recreation and residential play space for local children (Conway 1999, p. 29). This represented not only a shift in attitudes to religious spaces but also led to innovation in playground equipment and other facilities for entertaining and educating children, such as aviaries and petting zoos. Other sites such

24033 Platt Hall, Platt Fields, Manchester

Fig. 2.2 Promenading in Platt Fields park, postcard from Manchester in the early twentieth century

as those on the outskirts of towns were required to keep their natural beauty by their advocates and stakeholders. Similarly, Sefton Park in Liverpool opened in 1872 was designed to form part of the green belt around the city, its construction part-funded by housing development on axial streets overlooking the park (Historic England 2013). Its landscaping made the park look larger and more expansive by creating long vistas that gave views of church spires outside the park, connecting the town seamlessly with the park, through an undulating 'natural' valley design created by the Parisian-trained Edouard Andre (Conway 1999, p. 24). The expansion of Manchester's parks on the outskirts of the city centre was driven by the desire to save natural landscapes and open spaces, such as Boggart Hole Clough and Platt Fields, from housing development, as well as maximise space suitable for sports and for children's play: between the years 1890 and 1916 Manchester park land grew from 160 acres to 1400 acres, including large city parks, neighbourhood parks and inner city playgrounds (Latimer 1987, p. 13).

The budgets for new parks informed their design, but politics, existing land use, and a desire to demonstrate 'design thinking' were powerful motivators to push against budget constraints and business models. The making of Cheetham Park (also known as Elizabeth Park) in North Manchester demonstrates the difficulties of balancing the aspirations for serving the local population with political interests and land availability. The park was established 40 years after the initial municipal parkmaking flurry in the mid-century discussed above. The new park covered a rather modest 5 ½ acres, enclosed from its urban surroundings by brick wall and curb-mounted iron hurdles, and contained a host of features and amenities to ensure that the park might be used by locals as a space for exercise. A bowling green was provided alongside a boys and girls gymnasium, each one 'fitted up with the modern amenities'. The park also housed three lawns of roughly half an acre each and a series of paths criss-crossed the park, with tennis courts added in the fourth quadrant. There was a series of buildings erected on the site, including the entrance lodge, two shelters, a propagating house, tool-house and bowl-house. A news report of the park's opening day recorded that one of 'the primary features of the park is an extensive rockery which has been formed between the main entrance in Elizabeth Street and the bowling green' (Manchester Guardian, 26 September 1885).

This comprehensively equipped space was hard-won, and not without controversy. There was disagreement about an initial proposal to raise subscription for land purchase, which the Ratepayers committee overturned in favour of the Corporation of Manchester footing the bill, arguing that Cheetham ward had to date been unjustly ignored by the Parks Committee and deserved a suitable park as an act of moral duty. There was contention over the form the park should take, with one party wishing for green pastoral space in the middle of the red-brick housing which was rapidly taking over the area, close to the industrial units and warehouses of the textiles industry, and others recommending active recreation space:

> But I should not go to the expense of planting many trees or shrubs. What is really wanted is an open space for a playground, fitted up with all the necessary requirements. I would also suggest the erection on the upper part of this plot of a building similar to the Blackpool Winter Garden and Skating Rink, suitable for concerts, meetings, bazaars, exhibitions, &c., the revenue from which, if well farmed, would go a long way towards paying expenses. ("Cheetham", personal letter to *Manchester Guardian*, 27 August 1883)

This desire for rational recreation pre-empted the predominant policy rationales at the start of the twentieth century and incorporated an income model for sustaining the park's upkeep. The tensions between a park fit for promenading in nature and one that prompted active participation in team sports combined with the choice of land and difficult negotiations with aristocratic landowners. The options presented were a plot on an existing cricket field and the eventual site off Elizabeth Street owned by Lord Derby, which was closer to the residents of the area assumed to be in most need. There were protracted negotiations over price as the offer was reportedly less than half of the market value (MPCCMB, 26 February 1884) but the site was eventually secured for £9000. This was £1500 more than the larger cricket pitch plot, with additional costs for its infrastructure and amenities, and the deal was immediately criticised as poor business, as the site resembled a 'clay pit' in the middle of a housing estate, along with concerns that active sports would present a nuisance to the more refined residents.

THE POLITICAL CAPITAL OF PARKMAKING

Cheetham Park was created at the peak of municipal parkmaking in Manchester, and the debates over its site, contents and costs reflect the spatial and social mobility of class and status at this time. It also demonstrates the symbolic and political value that parkmaking has for the actors involved, a matter considered in this section. The many and varied features of Victorian parks, their bandstands, landscaping, buildings, sports equipment, playgrounds and the battles fought over their establishment mark out parks as places for civic pride and display not just of social etiquette but also ingenuity, horticulture and administrative expertise. Parks came in an era obsessed with the combination of science, technology and, to a lesser extent, art to produce private benefits through industry, and public good through philanthropy, in the absence of a welfare state. The parkmaking industry was no different.

As the section above describes, there were heterogeneous motivations for establishing public parks during the Victorian period, but significant evidence that they had, and continue to have, common non-partisan appeal and value in terms of political capital. Parks were difficult to refuse politically, or to make a case against, although there was some dissent to a universal acceptance of the value of parks in local rate-payers later in the 1920s and 1930s when attitudes towards investment in parks as municipal services conflicted with those who wanted to keep local rates down (O'Reilly 2019, p. 8). There was equally a popular understanding of how parks may have general private as well as public value, as O'Reilly reports, rate-payers lobbied for the funding of public parks through 'citizens committees' as they were "anxious to benefit from the increased land values that often resulted from the establishment of a park in a neighbourhood" (p. 9).

The policy drivers of healthy recreation, the formation of (middle-class) citizenship and the mechanisms of social control were persuasively packaged as public benefit, enshrined in the term 'people's parks' or 'parks for the people', which like the 'green lungs' became an important rhetorical device. This was evident in Queen's Park, one of the original three municipal parks in Manchester, which bore a plaque proclaiming: "this park was purchased by the people, was made for the people and was given to the people for their protection" (Conway 1985). Allan Ruff's history of the contemporaneous Philips Park describes how the opening ceremony speeches by the park's progenitor, Mark Philips, offered encouragement to the people of Manchester, via the huge procession gathered, to treat the park well, to value and care for it as their own, and to remonstrate with others who did not act the same. The city's mayor reportedly added more abruptly, "Mr Philips put me in mind that I should have told you, this park is now handed over to you and we are done with it" (Ruff 2016, p. 29).

The sentiment to this statement was not altruism, but pragmatic realism. The Whigs and Liberals of the new Town Council had inherited large debts and were without the powers from parliament to raise a levy on local rates and had to borrow and fundraise to make public improvements. Increasing local rates would not have endeared them further to the Tories and their supporters; therefore, a public campaign and subscription model were both politically and financially necessary. The campaign began by petitioning the mayor of Manchester for a meeting targeting city influencers, the "gentry, clergy, bankers, merchants and traders" (Ruff 2016, p. 10) for both their financial and vocal support. This meeting took place in August 1844, featuring rousing speeches from Mark Philips, pledges of funds and petition signatures, and successfully established a 25-strong Public Parks committee (p. 12–13). The arguments put forward for two or more public parks included the timeliness of action, since the town was in a better financial state than in the preceding years of 'cotton slump' (from the Mayor), and the need for accessible, proximal space for exercise and fresh air (from Philips). Philips also brought his own cosmopolitan class capital to the campaign, referring to a trip to Europe with Lord Egerton which he cited as the inspiration to create similar public spaces in Manchester as the public gardens he had seen in Rouen and Dresden.

A second meeting targeted the support of the working class in Manchester and Salford in September, attended by reportedly 5000–6000 people (Wyborn 1994). Ruff highlights the more complex arguments put forward at this meeting by the committee secretary, Tom Watts, to his fellow working men. These proposed the merits of studying and contemplating nature and made an allegory of the machine for the body, where fresh air oiled the joints, to advocate not just for the health benefits of bringing in green vegetation to the polluted streets of the city, but for their impact on social discourse:

> Does not the chemistry of nature present as large and pleasing field of study as the chemistry of tobacco smoke, gin and inflammation? May not conversation upon the growth and development of a plant, the rising and flowing of the life-fluid (the sap), the deposition of this fluid over the surface so as to form leaves, flowers and fruit, be made as enticing of the usual converse of the workshop, the street corner, the tap room. (Tom Watts, cited in Ruff 2016, p. 14)

He also added as an incentive that the wealthy had contributed, and so it was agreed to send subscription lists to workplaces and working men's committees, and the people's fundraising began. A year later, this was ceased when over £32,000 had been raised against an original target of £50,000, to add to a disappointingly small government grant of £3000. Subscriptions came from across social classes, with some later donors listed by factory in the local press (Layton-Jones 2016). The funds were enough to purchase three sites and take the three first municipal parks, Peel Park (Salford), Philips and Queens Park (Manchester) through design and build; however, no money was ring-fenced for maintenance, and suggested a more literal interpretation of the mayor's message to the people at the opening of Philips Park, cited above.

Public subscription was both a financial lever and symbolic gesture in other parkmaking projects, for example, Macclesfield's West Park, which followed Manchester's example four years later by raising £300 in "pennies from the poor" for the building of a park to commemorate Robert

Peel, in gratification for the repeal of the Corn Laws. In this case, the working men's efforts were enlisted first, with a calculated campaign designed by prolific fund-raiser, JP and chair of the Park Committee, John May,[5] which systematically canvassed workers at the many silk mills and factories in the town, resulting in 17,000 subscriptions in three weeks, before going out to the wealthier donors for funds (Anon 1888; Bentley Smith 2016). This level of subscription is arguably the greatest evidence of popular demand for parks, although there were likely ulterior motives for those leading the fundraising. The proposal of a park to provide suitable amusement for Macclesfield's lower classes, drawing on the example of the newly opened Manchester's zoological gardens and their perceived effect on local drunkenness, was noted in a speech to the Useful Knowledge Society in 1840 as turning "pleasure itself into a more powerful ally of virtue" (cited in Griffiths 2006). Macclesfield had 144 pubs and alehouses in 1850, the year the park was first mooted, and despite the town being surrounded by the open countryside of the Peaks and Cheshire plains, with its commons enclosed in 1804 and working hours of 70 hours a week leaving little leisure time for silk workers, there was a clear rationale for providing proximate green space for recreation in the town.[6] However, the establishment of this early park on the site of the 'Town Field' used as a temporary racecourse and for grazing out of season also signalled the collective desire to have the full range of civic institutions in place as a mark of being a proper, respectable town. West Park was not just for public welfare but to show it was not falling behind its nearby industrialised peers, such as Manchester, Salford and Derby (Griffiths 2006). West Park's landscape was designed by William Barron, who became known for axial design and new methods of mature tree planting, and who had been at Edinburgh Botanic Gardens before he became head gardener for Lord Carrington at Elvaston Castle, Derbyshire (Parkmasters 2007; Conway 2006). Amenities included an exhibition pavilion from a design by Augustus Pugin, an extremely large bowling green, a lodge, gymnasium, water fountain, men's urinal, shelters, an ancient market cross, two canons and a viewing mound (Anon 1888; Historic England 2001). The park's role in improving the town's social status was secured when it later acquired a museum (as discussed in Chap. 3) funded by Marianne and Peter Pownall Brocklehurst, siblings and children of local silk magnate and MP, to provide informal education and entertainment (Fig. 2.3).

Fig. 2.3 West Park, 1866 painting by George Stewart, schoolmaster of Macclesfield School of Arts. (Image courtesy Macclesfield Museums © The Silk Heritage Trust)

PUBLIC DISPLAY BY DESIGN

Town upmanship was unlikely to be the main motivation of those who provided the pennies for West Park, however. Far more common in park-making were donations from the wealthy, which came in a variety of forms: as cash donations, as gifted land and via (often negotiated) discounts to purchase prices. Alongside the other new civic institutions such as libraries, reading rooms, bathhouses, Sunday and technical schools, parks were powerful means to display and create political capital, as witnessed by the statuary, park furniture and amenities and the parks themselves which bear the names of their benefactors. The opportunity to showcase 'people's parks' philanthropy was also evident in their often lavish ceremonial openings. Elaborate proceedings, documented in accompanying published pamphlets and illustrated presentations, were choreographed to accommodate large assemblies and celebrate *en masse* the gift of a park to the town.

For example, the new Victoria Park in Macclesfield was donated by Francis Dicken Brocklehurst, cousin of Marianne, established on the grounds of Fence House, formerly part of the royal forest of Macclesfield, which Francis had inherited. The park opened on May 14 May 1894, acknowledged by flower garlands, evergreen decorations, flags and bunting strung across the streets and covering the businesses, civic buildings and shop premises within the town centres, the Union Jack flying from Town Hall and town's churches, which pealed their bells, and a day's public holiday for the town. A musical procession passed through the town featuring Friendly Societies and clubs and joined by aldermen, clergy, parade troupes in fancy dress, a Maypole on a tricycle, surrounded by "a bevy of young lasses" (Anon 1894, p. 23) and the May Queen on her throne. The ceremony in the park comprising speeches, presentation of the deeds, illuminated address and casket to Francis Brocklehurst was followed by a volley of gunfire, a firework display, gymnastic display, music in the bandstand and Maypole dancing. As the evening drew to night, there was a further firework display, including "a portrait of Mr. Brocklehurst in golden fire" (Anon 1894, p. 33) followed by illuminations at the front of the Town Hall using gas jets and coloured globes to project photographs of the park and a "speaking" image of the park's donor onto the building.[7] The opening ceremony for a park was a key opportunity, therefore, to address and thank founders and funders, acknowledge the importance of local civic infrastructure and national political figures, such as Robert Peel, to show respect for military forces and campaigns, and allegiance to the monarchy, commemorating anniversaries, birthdays and other national events.

Whilst parks provided political capital for local philanthropists, aldermen and collectively for the town, they also provided the mechanism for professional development and promotion of those other parkmakers, the landscape gardeners and horticulturalists who became the architects of nature in the urban context, heralding new movements of landscape design. Park historians have highlighted the men (and it was predominantly male-dominated) who led the field, with special mentions in particular for the influence of Humphrey Repton, John Nash, John Claudius Loudon and Joseph Paxton, who brought their skills from patronage for private estates to public commissions in the early Victorian parks (Conway 1996; O'Reilly 2019).

The stories of these individuals focus on their industriousness, technical skill and tendency towards self-promotion, often through publication of journals and textbooks on their preferred design and horticultural methodologies and innovations. They were influential not just in their own spheres but on policy transfer, as park schemes and horticultural techniques were imported and adopted in other cities and countries. Landscaping, layout, selection and signposting of facilities reflected the skills and approaches of the designers, drawn on and extended by subsequent park planners. Joseph Paxton's lack of incorporated sports spaces in his plan for Birkenhead Park on Merseyside was remedied in Joshua Major's plans for the Manchester parks, whilst his body of work in Glasgow,

Halifax and London (most notably Crystal Palace Park) became highly influential for many parks and pleasure gardens (Conway 1996). His designs influenced Frederick Olmsted who visited Birkenhead Park and incorporated them in his making of New York's Central Park (Elborough 2016; Conway 1996).

Another major influence on Olmsted and the blueprint of the public park was found in John Claudius Loudon, who designed the early semi-public Derby Arboretum in 1840. Also claimed as the first municipal park in England, although technically founded by local philanthropists, the park was closed to the general public for free access for five days a week for its first 40 years, in part to ensure the costs of its upkeep were low, but also to maintain order and protect its contents. The Arboretum featured highly specialised landscaping and planting, displaying rare botanical specimens as a "living museum". The land was donated by local textile manufacturer, Joseph Strutt, with members of his family and other associates who were part of Derby's radical bourgeoisie, and who like Loudon, were influenced by both Charles Darwin and Jeremy Bentham in their philosophical and philanthropic pursuits, believing in progressive scientific education and rational recreation for social reform (Elliott 2001). Loudon was a prolific writer and editor of three contemporary magazines in gardening, natural history and architecture, which did much to publicise his philosophies and promote education on these topics to their middle-class readership. He lobbied for open space for public recreation, influenced by his observation of changing conditions in London, proposing circular 'green belts' and promenades leading out in concentric circles from urban centres, which were blueprints for garden cities (Jefferson 2019).

Loudon was just one of a number of star parkmakers whose professions were built on the patronage of the landed gentry; however, the design of the public park was less homogeneous than is suggested by a focus of park historians on flagship parks and their makers. As parkmaking continued at pace into the twentieth century, policy rationales were dominated by the need for spaces for physical recreation to train and grow a healthy population linked to military supply ahead of the First World War, which shaped the playing fields movement and ensured parkmaking and management remained a key feature of town planning (Conway 1996). Conway recognises that the majority of public parks in this period were not the product of showmen such as Paxton, Barron, Loudon and Thomas Mawson, who became a leading park designer in the first quarter of the new century. Instead, they were designed by the engineers, surveyors and

superintendents under local authority employ who brought new managerial approaches to the use and maintenance of landscape design. She sin-
gles out three particular superintendents for recognition: J. McHattie in
charge of Edinburgh Parks; Lieutenant-Colonel J. J. Sexby, a major figure
in turn-of-the-century park management in London, overseeing a team of
1000, and publishing his book on London parks in 1895; and William
Pettigrew, who presided over Manchester parks from 1915 to 1932 and
who wrote the first handbook on park administration in 1937 (Conway
1996, p. 37).

These, along with many other unnamed administrators, shaped the policy environment of parks through their establishment of administrative and
regulatory processes and negotiation and management of budgets, as well
as through their influence on spatial design, cultural programming and
community engagement. Furthermore, the public-ness or otherwise of
municipal parks, and their success as spaces for reform, moral improvement
and regulation of the social body was dependent not only on their design
but also their policy environments, wider political and economic conditions
and on the publics who used and participated in them. In the final section,
I consider the changing environments for municipal public parks in the
twentieth century which lay the basis for their contemporary challenges.

THE DECLINE OF THE PUBLIC PARK

Historical research has paid more attention to parkmaking during the
nineteenth century than in subsequent decades (O'Reilly 2019). The
studies that have been undertaken suggest that whilst the imperatives for
parkmaking continue to highlight the need for a resolution of the urban
problem, through spaces for recreation to support the health of the nation,
what characterises parkmaking in the twentieth century is its institution
into town planning and the everyday landscape of urban locations (Conway
2000). O'Reilly notes the shift in policy rationale for these parks from the
function of moral improvement and regulation to that of public service,
offering safe play and recreation for families and children within their
neighbourhoods (O'Reilly 2013). In inner city areas, smaller play and
pocket parks maintained the Edwardian commitment to provide parks
where possible on available space.

Parks were a site for active citizenship, collective regulation and the performance of gender relations in the context of the changing landscape of
the early twentieth century, for example, in the use of parks for political

demonstration within the women's suffrage movement (O'Reilly 2009). The years between the wars saw a focus on large playing fields, funding through new sources such as the George V Memorial Fund, incorporating traditional 'park' features such as gardens, bandstands and bowling greens, and the take-up of new methodologies that linked public green spaces across locales through chains and parkways, designed in consultation with housing and roads departments. For example, the Princess Parkway in Manchester, opened in 1932, was designed to be a gateway to the garden suburb of Wythenshawe, rather than a link road from the centre to the motorway network which it later became (Wythenshawe History Group n.d.). Park horticultural techniques crept into this new green infrastructure, such as flower beds on roundabouts and shrub planting on traffic islands, and there was removal of railings from existing parks not just to support the war effort but to create more porosity between green and grey space.

Wartime and mid-war uses for public parks were varied and prolific. They ranged from sports training for a healthy, disciplined youth for conscription to the supply of metal, land for food growing, with flowerbeds turned into allotments and park amenities turned over to animal husbandry. Parks played a role in morale boosting and solace, from play spaces for evacuees and outdoor concerts, as well as later commemoration of the services, the wounded and the dead. Land was also requisitioned for defence training and resources: trenches were dug, grass used for sandbags, and buildings in parks housed decanted administrative departments. For example, Platt Hall in Platt Fields Park, Manchester, was the home for war refugees from Belgium and then conscientious objectors, during the First World War, and the administrative headquarters for the Manchester police, then Manchester School of Art during the Second World War (Manchester Art Gallery n.d.).

Conway (2000) traces the development of post-war parks, exploring the design shifts in reconstruction and the influence of modernism on creating 'green deserts' of landscaping with clean lines and low maintenance. Design innovation combined with post-war morale boosting, for example, the influence of the Festival of Britain in 1951 on pleasure grounds, through colourful planting and spaces for fun, relaxation and amusement, led to new additions to regional parks such as new gardens and illuminations. These provided a counterpoint to the investment in spaces for commemoration contemplation that dominated the period through the establishment of monuments, peace gardens and children's play areas. Legislation giving powers to local authorities to rebuild towns and cities included the National Parks Act 1948, which led to the establishment of country parks, later extended by the Countryside Act of 1968. The decentralisation of towns through the development of out-of-town shopping malls and industrial units saw the urban park eclipsed by the creation of country parks on the edge of towns, whilst the garden cities movement embedded parks and green space in whole city plans. Influenced by the neighbourhood planning theory of Clarence Perry, these new towns not only attempted to standardise quotas for the amount of green space retained in urban development but, in some cases such as Milton Keynes and Telford, used parks and 'desire lines' across open green spaces as the starting point for their layout (Conway 2000).

Over the latter half of the twentieth century, there were significant changes in the way that parks were treated as policy spaces, linked to the increased resources and choices people have for their greater disposable income and leisure time (Lamond 2019). The location of park administration shifted in 1967 following the Maud Committee on local government and subsequent Bains Report which attached parks to local leisure and events strategies, losing them their relative independence (Layton-Jones 2016). They were integrated into leisure departments through the Local Government Act of 1972, a move that aimed to improve efficiency but led to an increasing detachment from localised skills and revenue management within individual sites. Furthermore, it led to the privatisation and contracting out of public services, as local authorities moved to competitive tendering, leaving parks vulnerable to the political struggles of the mid-1970s between local and central government (Crowe 2018). Increasing access to culture and leisure through mediated and commercial means, and the privatisation and domestication of sports and recreation, exacerbated the competition with other non-statutory services for investment at a local authority level, in a climate of year-on-year cuts, as discussed further in Chap. 5. Parks became neglected by their publics and their owners.

Conclusion

This chapter has presented an overview of parkmaking history, with a focus on the English towns and cities that were first to test out models for cultural management through the provision of open green spaces, in close enough proximity to serve working populations. It examined the range of rationales for fundraising and investment that circulated amongst the Victorian political classes, stowed by the discourses of scientific progress, biotic metaphor and design thinking and expressing the dominant cultural norms and values associated with class formation and distinction, social reform and improvement. Significantly, it notes the absence of everyday park-users within the historiography of public parks, reflecting their lack of access to the cultural public sphere (McGuigan 1992) through which the archive and historical record is composed, despite their role as recipients of parks as policy spaces and, as I argue in the following chapters, participants in the practising of public space. Along with other new civic and cultural institutions, the making of public parks in the nineteenth century was not only an important strategy for social and moral improvement, managing public space and urban populations but also a fundamental form of municipalisation which saw the responsibility of the public park become that of local government. However, the political economy of parkmaking created business models which, as a legacy of their Victorian heritage, serve to undermine their sustainability within changing municipal policy environments and contemporary contexts of "creeping neoliberalism" (Crowe 2018), as discussed further in the chapters that follow.

Notes

1. For example, the seven 'hays' or enclosures mentioned in Domesday in the Macclesfield Manor included a mediaeval deer pound and administrative centre at Toot Hill at the north east top of Macclesfield Forest. Pre-Conquest hunting grounds under Earl Edwin, this royal forest was later favoured by the Black Prince, son of Edward III (Hey 2014). The area became known in the fourteenth and fifteenth centuries for the longbow skills of the Cheshire Archers, elite soldiers who were recruited from the Macclesfield Hundred as royal bodyguards and into service in battle during the Hundred Years War. Training in the skills of archery was an expectation for every man, both commoner and nobility, enshrined in law under Edward III to ensure the supply of archers. Evidence of the proliferation of archery is found commonly in place and street names, including the term "Butt" or "Butts," indicating the places for target practice, which was compulsory by law on every Sunday from 1363. Indeed, archery was not only a source of military supply but also of revenue and potential social mobility for the common man and was seen as morally preferable to other contemporary leisure pursuits such as card-playing and bowling (Loades 2013).
2. William Emes mentored John Webb, who was responsible for building a 4-mile wall to enclose Heaton Park, and who may also have been responsible for Wythenshawe park (Manchester City Art Galleries 1987).

3. The intransigence of this popular belief in miasmas can also be witnessed in the epidemiology of cholera outbreaks and the work of John Snow who advanced the cause significantly by proposing that the disease was not air-borne but water-borne, a suggestion met with wide resistance (Thomas 2015).

4. Prostitution was similarly mapped in cities such as Liverpool, and police statistics and records of violent and property crimes and the appearance and physiology of their perpetrators fill local archives, complementing the more vivid accounts reported in the press.

5. John May also championed workers' holidays to Blackpool, which tradition-ally took place during Barnaby, the silk workers' fortnight around the former charter fair day in June. He had previously led the campaign to open the public baths in the town. It is notable that West Park (first named Peel Park) opened during wakes week in October, allowing the townspeople and many other out-of-town visitors to attend (Anonymous, 2011; Historic England 2001).

6. Griffiths notes a number of ways in which West Park's performance against policy objectives was measured after its establishment, including the number of visitors and donations of objects, the use of planting and care for flowers for education and moral improvement, and the collection of statistics on reported decreases in drunkenness and local crime associated with the park's opening the year before (Griffiths 2006: 2016).

7. Despite this intricate homage to the philanthropist, Griffiths finds that the evidence weighs in favour of altruism away from outright political instru-mentalism in both the case of Victoria Park and in the later establishment of South Park by William Frost, a silk manufacturer in Macclesfield, who shunned a large public ceremony, instead opening the park quietly on a weekday in 1926. Both appear to have given for public benefit rather than private reward through political advancement, although she notes that this was not the case for other donors who were recognised by park statuary and buildings (Griffiths 2006).

REFERENCES

Anonymous. 1894. *Record of the proceedings at the opening of the new Victoria Park*. Macclesfield: Claye, Brown and Claye, Courier Office.

———. 2011. *A* Walk through the public institutions of Macclesfield, being a series of articles...reprinted from the Macclesfield Courier and Herald, 1888. British Library Historical Reprint. London: British Library.

Arnold, M. [1869] 1981. *Culture and anarchy*, ed. J.D. Wilson. Cambridge: Cambridge University Press.

Barker, H. 2004. 'Smoke Cities': Northern Industrial Towns in Late Georgian England. *Urban History* 31 (2): 175–190.

Bennett, O. 2005. Beyond machinery: The cultural policies of Matthew Arnold. *History of Political Economy* 37 (3): 455–482.

Bennett, T. 1995. *The Birth of the Museum*. London: Routledge.

Bentley Smith, D. 2016. *Past times of Macclesfield*. Vol. III. Stroud: Amberley Publishing Limited.

Booth, N., D. Churchill, A. Barker, and A. Crawford. 2020. Spaces apart: Public parks and the differentiation of space in Leeds, 1850–1914. *Urban History* 48: 552–571. https://doi.org/10.1017/S0963926820000449.

Briggs, A. 1963. *Victorian Cities*. London: Odhams Books.

Bullock, T.A. 1857. *Bradshaw's illustrated guide to Manchester*. London: W.J. Adams.

Chetham's Library. 2013. M-M-M-M-M-M-M-My Pomona. https://library.chethams.com/blog/m-m-m-m-m-m-m-my-pomona/. Accessed 6 May 2023.

Conway, H. 1985. The Manchester/Salford parks: Their design and development. *The Journal of Garden History* 5 (3): 221–260. https://doi.org/10.108 0/01445170.1985.10410540.

———. 1996. *Public parks*. Princes Risborough: Shire Publications Ltd.

———. 2000. Everyday landscapes: Public parks from 1930 to 2000. *Garden History* 28 (1): 117–134. https://doi.org/10.2307/1587123.

Crompton, J.L. 2017. Evolution of the "parks as lungs" metaphor: Is it still relevant? *World Leisure Journal, 2017* 59 (2): 105–123. https://doi.org/10.108 0/16078055.2016.1211171.

Crowe, L. 2018. The future of public parks in England: Policy tensions in funding, management and governance. *People, Place and Policy* 12/2: 58–71.

Dreher, N.H. 1993. Public parks in urban Britain, 1870–1920: Creating a new public culture. Publicly Accessible Penn Dissertations. 2673. https://repository.upenn.edu/edissertations/2673

Elborough, T. 2016. *A Walk in the Park: The Life and Times of a People's Institution*. Jonathan Cape.

Elliott, P.A. 2001. The Derby arboretum (1840): The first specially designed municipal Public Park in Britain. *Midland History* 26 (1): 144–176. https://doi.org/10.1179/mdh.2001.26.1.144.

Gilmore, A. 2013. Cold spots crap towns and cultural deserts: The role of place and geography in cultural participation and creative place-making. *Cultural Trends* 22 (2): 86–96. https://doi.org/10.1080/09548963.2013.783174.

Gilmore, A., and P. Doyle. 2019. Histories of public parks in Manchester and Salford and their role in cultural policies for everyday participation. In *Histories of cultural participation, values and governance. New directions in cultural policy research*, ed. E. Belfiore and L. Gibson, 129–152. London: Palgrave Macmillan.

Gray, C. 2002. *Local Government and the Arts Local Government Studies* 28 (1): 77–90. https://doi.org/10.1080/714004133.

Griffiths, S. 2006. The charitable work of the Macclesfield silk manufacturers, 1750–1900. Unpublished doctoral dissertation, University of Liverpool.

Gunn, S. 2007. *The public culture of the Victorian middle class: ritual and authority in the English industrial city 1840–1914.* Manchester: Manchester University Press.

Hall, P. 1996. *Cities of tomorrow: Updated edition.* Oxford: Blackwell.

Haslam, D. 1999. *Manchester, England: The story of the pop cult city.* London: Fourth Estate.

HCPP/House of Commons Parliamentary Papers. 1833. Report from the Select Committee on Public Walks; with the minutes of evidence taken before them, XV.337 Vol. 15, Paper 448. http://parlipapers.proquest.com/parlipapers/docview/t70.d75.1833-014187?accountid=12253. Accessed 15 Apr 2023.

Hey, D. 2014. *The history of the Peak District moors.* Barnsley: Pen & Sword Local Ltd.

Hill, K. 2005. *Culture and class in English public museums, 1850–1914.* Aldershot: Ashgate.

Historic England. 2013. Sefton Park, amended September 2013. https://historicengland.org.UK/listing/the-list/list-entry/1000999#:~:text=Sefton%20Park%20was%20formed%20from,79)%2C%20a%20Liverpool%20architect. Accessed 15 Apr 2023.

———. 2001. Listing of West Park, park and gardens Grade II listing. https://historicengland.org.uk/listing/the-list/list-entry/1001495.   Accessed 11 Aug 2020.

———. 1986. Wythenshawe Park, Official List Entry, 20 February 1986. https://historicengland.org.uk/listing/the-list/list-entry/1000857?section=official-list-entry. Accessed 20 Dec 2023.

Howkins, A. 2011. The commons, enclosure and radical histories. In *Structures and transformations in modern British history,* ed. D. Feldman and J. Lawrence, 118–141. Cambridge: Cambridge University Press.

Jefferson, E. 2019. Park life: On John Claudius Loudon, the father of the modern park. *CityMetric,* March 22. The New Statesman. https://www.citymetric.com/fabric/park-life-john-claudius-loudon-father-modern-park-4532. Accessed 20 Apr 2023.

Jerram, L. 2011. *Streetlife: The Untold History of Europe's Twentieth Century.* Oxford: Oxford University Press.

Jones, K. 2022. Green lungs and green liberty: The modern city park and public health in an urban metabolic landscape. *Social History of Medicine* 35 (4): 1200–1222. https://doi-org.manchester.idm.oclc.org/10.1093/shm/hkac055.

Joyce, P. 2003. *The Rule of Freedom: Liberalism and the Modern City.* London: Verso.

Lamond, I. 2019. Leisure as an object of governmental policy in UK elections: 1945 to 1983. *Leisure/Loisir* 43 (2): 249–269. https://doi.org/10.1080/14927713.2019.1613170.

Latimer, C. 1987. *Parks for the people, booklet for exhibition 'parks for the people: Manchester and its parks', 1846–1926*. Manchester: Manchester City Art Galleries.

Layton-Jones, K. 2016. History of public park funding and management (1820–2010). Historic England, 2016, 135.

Loades, M. 2013. *The Longbow*. Osprey Publishing, Oxford.

Manchester Art Gallery. n.d. Platt Hall. https://manchesterartgallery.org/visit/platt-hall/. Accessed 17 Aug 2020.

McGuigan, J. 1992. *Cultural Populism*. London: Routledge.

Olechnowicz, A. 2000. Civic leadership and education for democracy: The Simons and the Wythenshawe estate. *Contemporary British History* 14 (1): 3–26. https://doi.org/10.1080/13619460008581569.

O'Reilly, C. 2009. Women in Manchester's Edwardian parks 1900–1935. In *Women's life and leisure in the twentieth century*, 25/11/09. Staffordshire University. (Unpublished).

———. 2013. From 'the people' to 'the citizen': The emergence of the Edwardian municipal park in Manchester, 1902–1912. *Urban History* 40 (1): 136–155. https://doi.org/10.1017/S0963926812000673.

———. 2019. *The greening of the city: Urban parks and public leisure, 1840–1939*. Vol. 73. London: Routledge.

Parkmasters. 2007. William Barron and West Park, Macclesfield, 03:45 22/07/2019, BBC Radio 4 Extra, 15 mins. https://learningonscreen.ac.uk/ondemand/index.php/prog/140C65CD?bcast=129758100.   Accessed 11 Aug 2020.

Powell, M., and T. Wyke. 2015. Counting the coppers: John Jennison and the Belle Vue Zoological Gardens. In *Culture in Manchester: Institutions and urban change since 1850*, ed. Janet Wolff and Mike Savage, 34–60. (Manchester, 2013; online edn, Manchester scholarship online, 22 Jan 2015). https://doi-org.manchester.idm.oclc.org/10.7228/manchester/9780719090387.003.0003. Accessed 6 May 2023.

Rees Leahy, H. 2012. *Museum bodies: The politics and practices of visiting and viewing*. Farnham: Ashgate.

Ruff, A. (2016). *Manchester's Philips Park: A Park for the People, By the People, Since 1845*. Stroud: Amberley Publishing.

Russell, C. 1905. *Manchester boys: Sketches of Manchester lads at work and play*. Manchester: Manchester University Press.

Sigsworth, M., and Worboys, M. 1994. The public's view of public health in mid-Victorian Britain. *Urban History* 21 (2) (October 1994): 237–250.

Solnit, R. 2000. *Wanderlust: A history of walking*. New York: Viking.

Spiers, M. 1976. *Victoria Park, Manchester: A nineteenth-century suburb in its social and administrative context*. Manchester: Manchester University Press for the Chetham Society, 1976. Print.

Standing, G. 2019. *Plunder of the Commons*. Penguin Books
Thomas, A. 2015. *Cholera: The Victorian plague*. Havertown: Pen and Sword.
Wyborn, T. 1994. *Parks for the People: the Development of Public Parks in Manchester, c1830–1860*. Manchester: University of Manchester.
Wythenshawe History Group. n.d. Wythenshawe's history and heritage. http://www.wythenshawe.btck.co.uk/DownMemoryLane/PrincessPark. Accessed 17 Aug 2020.

CHAPTER 3

Pleasure Grounds and People's Palaces: The Museum in the Park

Abstract This chapter explores the histories and practices shared by public parks and their co-located museums, firstly, during a turbulent period of class formation and municipalisation in the mid-to-late nineteenth century, and, secondly, within the contemporary context through case studies of museum-making in industrial towns and cities in the north of England. The motivations for the acquisition and display of artworks and collections within these museums share provenance with the philosophies of nineteenth-century parkmaking, rooted in social reform, education and improvement, and allied to class formation and place governance. The chapter explores how the 'art reformers' of regional and provincial towns and cities formed literate cultural public spheres advocating for new museum spaces co-located within new municipal parks. The legacy of these museum-makers continues within the practices of contemporary museums in parks as they attempt to encourage publics, through community engagement and social practice, to cross museum thresholds using park spaces as contact zones and draw in those outside the museum walls.

Keywords Municipalisation • Co-located museums • Art reform • Cultural public sphere • The gallery in the park • Threshold fear • Community engagement • Contact zones • Ruskin

A. Gilmore, *Culture, Participation and Policy in the Municipal Public Park*, Palgrave Studies in Cultural Participation, https://doi.org/10.1007/978-3-031-44277-3_3

67

INTRODUCTION

In her book on the politics of popular protest in eighteenth and early nineteenth-century Northern England, Navickas (2015) discusses the stretching and spatialisation of the Habermasian concept of the public sphere by fellow historians of this period. Whilst Habermas allocated the public sphere as an arena of bourgeois power of the 'long eighteenth century' comprising elite urban spaces such as coffee houses and the discursive spaces mediated by the newspapers and printed pamphlets, these historians identified counter-public spheres within working-class movements of protest, the sites they occupied and practices they incorporated. The physical spaces for assembly and debate within the industrial urban fabric, such as the street and the pub, were augmented by ritual practices such as processions and parades; Navickas explores the drawing and redrawing of boundaries between public and private spaces as dissenters and reformers navigated the surveillance and sanctuary of new civic buildings and organisations, including assembly rooms, chapels and radical societies. She charts the development of these spatialised strategies of resistance, domination and hegemony, and their displacement and absorption as the long eighteenth century progressed.

In this chapter, I explore the co-location of cultural buildings, programming and communities in the municipal public park and the ensuing practising of public space as a form of spatialised public sphere making (and at times subversion and disruption). In the book *The Structural Transformations of the Public Sphere*, Habermas (1992) discusses the question of representative publicness as the performance of being a public person, involving practices which lead to the connotation of status, legitimacy and entry into and membership of civic society. He identifies emerging distinctions between the aspiring bourgeoisie and declining aristocracy in eighteenth-century Germany which underpin the practising of publicness through rational discourse between equals, replacing the hierarchical structure of sovereignty that is articulated through political representation (Pan 2014).

Habermas sets his origin story for the public sphere within the context of changing patterns of land ownership, the decline of feudalism and development of a (to some extent) coterminous 'civil society' which is autonomous from corporate, state and church ties. These conditions laid the foundation for the explosion of print media which carried forth into the world new forms of representation in the form of "news" to the "reading public" of private, commercial and mercantile capitalism (1992, p. 171). As Fraser (1990) observes, this identification makes visible the structural space within which there is the possibility for participatory democracy:

> It is the space in which citizens deliberate about their common affairs, and hence an institutionalized arena of discursive interaction. This arena is conceptually distinct from the state; it is a site for the production and circulation of discourses that can in principle be critical of the state. The public sphere in Habermas's sense is also conceptually distinct from the official economy; it is not an arena of market relations but rather one of discursive relations, a theater [*sic*] for debating and deliberating rather than for buying and selling. (p. 57)

This possibility is qualified by Fraser's identification of power differentials between "weak" and "strong" publics: to deliberate might be solely to form opinion rather than enact change, and so requires other foundations for political action (Fraser 1990). Nevertheless, this conception reaffirms the public sphere as one of symbolic and ideological (rather than economic) exchange which stands between the state and civil society (Barrett 2010). It is not and never culturally neutral or devoid of power differentials articulated by the expression of cultural values, which ferment and cement a political economy that in turn amplifies the structural relationships between unequally empowered social groups (McCarthy 1992). This is not simply a fluid form of discursive or symbolic power but rather one that is embodied and practised, that inhabits spaces and informs the spatial characteristics of places and their material forms.

As described in Chap. 1, the significance of the Habermasian public sphere in rethinking the logics of cultural policy is recognised by McGuigan (1992, 2004), who argues that the affective communications of the cultural public sphere can resist the commodification and appropriation of power and agency by private corporations. The public sphere conceived by Habermas foregrounded literary discourses and practices rather than those of broader cultural institutions and forms, as discussed by Barrett (2010) as she applies his ideas in her consideration of the museum as a public sphere. The literary public sphere requires a 'reading public' to function; therefore, translation to more inclusive forms of cultural expression and means of practising participation has democratic potential. Barrett points out that Habermas does not prescribe specific material spaces for the operation of the public sphere although he does refer to coffee houses and markets within the context of its origin. The proposition of the museum as a material space for practising publicness beggars an evaluation of museums as spaces of democracy, and Barrett considers that when applied conceptually to museums, their spaces and publics, the public sphere can be a diagnostic for understanding how these cultural institutions stand in the gap between government and civil society. We can see what kinds of public spaces are generated and practised, explore whether the relationships between museums, their publics and the state are strong or weak, conducive or otherwise, to affective communications, resistance and agency, and understand the significance of cultural institutions for participation and democracy.

This book investigates municipal public parks as part of urban policy infrastructures that shape and govern everyday culture. In Chap. 2, I considered the history of parkmaking in England and drew attention to the policy rationales for municipal investment in public parks that followed the interests, norms and values of urban elites. The early municipal parks of the nineteenth century were intended as spaces for public walking and display, for social mixing, rational recreation and healthy pursuits away from those that might interfere with the governance of place and the interests of capitalism. In addition, both then and now, parks are important sites for the public provision of culture with a big 'C', through open-air music concerts, outdoor theatre, circus arts and festivals and as the landscaped environment for artworks, sculpture and statuary in the public realm. They are the sites of cultural venues, such as museums, libraries, cinemas and their associated cafes and restaurants, education programmes, gardening and volunteer projects, alongside myriad other forms of cultural participation and social assembly. Parks are cultural spaces that connect to and promulgate other forms of cultural institution and creative practice within urban public and proto-public realms, embedded in the power complexes and relational networks of local cultural ecosystems and in competition (and sometimes collaboration) with other publicly funded and owned cultural institutions. Like Barrett's museums, this affords the opportunity to consider how participation practices and discourses surrounding parks and their co-located institutions enmesh as affective communication which helps us to collectively make sense of the world, connote status, demarcate social groups, and formulate strategies of resistance within public space, as a cultural public sphere.

This chapter begins by considering parallel histories of municipal public parks and museums, highlighting their synchronous expansion and consolidation in the nineteenth century, and the synergies and tensions between them that remain today. Like parks, museums and galleries were advanced for their benefits to the public and moral health of labouring populations and the need to ensure social control through reform of tastes and practices of participation. Through archival and secondary research and expert witness interviews,[1] the chapter traces the roots of socially engaged museum practice back to the location of the municipal park, and the influence of nineteenth-century English cultural intermediary, John Ruskin and his lesser-known followers and successors, Thomas Horsfall and Marianne Brocklehurst. These protagonists put into practice their ideas combining the utility of education in the arts and in immersion in nature as a strategy for the improvement and welfare of the working classes, creating new institutions and practices which, I argue, materialise and transform the cultural public sphere and remain reference points for the ambitions and missions of cultural institutions and their sponsors today.

The first section introduces the art gallery in the park within the broader context of Victorian museum making. We go back to Ruskin to consider his interest in arts education within parks through the co-location of museums, remaining in the North of England to consider Ruskin's endeavours in the city of Sheffield and his influence on Thomas Horsfall's Manchester Art Museum, which was founded in Queens Park in Manchester before moving to the slum neighbourhood of Ancoats, and which in turn influenced the establishment of West Park Museum by Marianne Brocklehurst in Macclesfield. These examples of museum-making are contingent with parkmaking, and I consider how the rationales for their design, curation and sponsorship presented in the vernacular spaces of these bourgeois public spheres, the committee meetings, correspondence and print media coverage, reveal the discursive relations and network ties of local philanthropists, industrialists and do-gooders involved.

The chapter tracks the influence of these protagonists forward to contemporary examples of museums and galleries co-located in parks in Manchester and Salford and which use their outdoor spaces for community engagement, co-production and advocacy (Alford 2008; Kershaw et al. 2018). These examples reveal the missions of those with stakes in their material spaces, the natural environments of museums in parks, and within the policy environments, the local cultural ecosystems through which they are constituted and governed. Museums in parks hold the democratic potential of their environments as contact zones (Clifford 1997) for social encounter, audience development and locational citizenship (Di Masso 2015) but also obstacles to participation such as physical thresholds and social boundaries, the enclosures of cultural taste, value and literacy presented through museum curatorial and management strategies. The chapter concludes by considering the opportunity for a counterpublic sphere which engages museological strategies for practising publicness outside the walls of the museum, within the park.

MUNICIPALISATION, MUSEUMS AND PARKS

The co-location of art galleries and museums in English municipal public parks took root both materially and conceptually in the mid-to-late nineteenth century, the product of a continuing evolution of cultural policies that instigated new practices of reform and regulation of the public body through the promotion of specific forms of cultural participation. In Europe and America, throughout the nineteenth century, the urban fabric was being transformed by a proliferation of new spaces which brought state and society into contact, replacing the street as the previous location for this encounter and planning public order into city design (Barrett 2010). Municipal museum-making gathered pace under the umbrella of instructive or rational recreation, which included the formation of 'settlements' and urban 'missions' to instruct and educate people while encouraging social mixing and productive pastimes. As Waterfield (1994) describes, such social initiatives were intended to recolonise working-class spaces characterised by population churn, overcrowding and compressed leisure time in newly industrialised cities. There was mass building, renovation and extension of town halls, assembly rooms, workhouses, penal institutes, hospitals, free libraries, Sunday schools and venues for philanthropic and cooperative educational programmes, such as Mechanics Institutes and Friendly Societies, alongside new infrastructure for transport, sanitation and improvements to public realm (Hill 2005; Anon 2011). New buildings designed for specific types of arts and culture participation, such as museums (Forgan 2005) and concert halls (Scott 2002) joined a proliferation of civic buildings, employing a new tier of local specialist architects (Black and Pepper 2012) and mostly initiated by private interest before being transferred to or supported by municipal local authorities.

The buildings for these new institutions were often eclectic in style, design and the practices that they promoted, although there was some standardisation and regulation which arguably led to institutional isomorphism as the century progressed. The legislative framework that permitted fundraising from the public at a local level was transformed by a series of parliamentary bills to levy funds for museums and galleries in 1845, joined by libraries in 1850, through the Public Libraries Act. Municipal authorities with populations greater than 10,000 could raise up to one-halfpenny in the pound through taxation to be spent on capital and revenue costs. This built on the 1835 Municipal Corporation Act, which opened up local government and brought into being a new municipal elite, creating the drive for local reform although requiring national legislation to realise the resources without dependence on philanthropy or private interest (Hill 2005).

The development of municipal galleries and museums was slow and uneven, dependent on local support and the capacity to raise funds, whether private or public. However, those that were established proved popular for visitors and encouraged participation in the creation and display of collections in regional urban centres through the patronage of their developing middle classes. Far less dependent on acquisition of major works from contemporary artists than on gifts and bequests, they provided opportunities for the "established bourgeois culture: a generally declining aristocracy played no part in their formation, any more did the working class" (Waterfield 1994, p. 35). With the support of the middle-class lobby, these museums became pragmatic and affordable mechanisms to signal class capital and legitimise authority, whilst creating new arenas for their own pursuits (Hill 2005).

Like parks, museums were instruments of social control within turbulent times of class formation and social re-ordering, bound into places through the visible improvement and transformation of city spaces and the public realm. They similarly required behaviour that took effect on the public body (Rees Leahy 2012), demanding specific dress and comportment. Unlike parks, their enclosure of public space within imposing buildings presented further barriers to working-class participants, requiring sufficient interest and enthusiasm to cross their thresholds and go inside. Nevertheless, their establishment brought new visitor destinations to many industrialising cities and towns in countries across Europe, North America and Australia in the nineteenth century. This was no simple case of policy transfer, nor was it a coherent strategy produced by a unified middle class. Rather, museums and their constituents were sites of continual battles between groups with different visions of art's purpose and how best to institute this into the messy complexities of urban management. The creation of parks and museums as civic spaces for social reform was not merely parallel but intimately connected. Both required resources for development, architectural and landscape design, local advocates to promote and bankroll these new buildings and public realm, and organisational strategies for engaging and managing publics intended to be the beneficiaries of social improvement. The Victorian social commentator and art critic John Ruskin (whom we met first in Chap. 1) was pivotal in bringing these two spaces together through the cultural framework developed in his public lectures, letters and writing and taken up by interested elites who could similarly afford to pursue his ambitions. Known as Ruskinians, these Victorian fanboys (and occasional fangirls) established local chapters to take forward his teachings into local cultural policy within their cities and towns. Ruskin's influence in this way on the aspiring second cities of Birmingham, Liverpool and Manchester is explored by Woodson-Boulton (2007). She tells the story of these 'art reformers', who by their advocacy for museum-making:

> overcame long-standing British aversion to taxation, and even longer-standing suspicion of art as foreign, idolatrous, and aristocratic, to make cities provide art to their citizens. In turn, however, governments transformed the meaning of art, putting it on the municipal payroll in order to proclaim the achievement of local self-rule and the success of the industrial city. (p. 52)

The transformation that Woodson-Boulton describes did not just expand the meanings of art but also the purposes to which it was put within the practices and spaces of moral improvement, and in turn its function for the making of polity. Art was seen to have unique powers for social healing, moral and cultural regeneration, design and technical training, as well as to "raise the civic profile and provide a key arena for genteel sociability" (Woodson-Boulton 2007, p. 49). As a gatekeeper to participation in arts education within museums, municipal governments could display their local autonomy and political capital, intervening in the lives of the urban poor and taking responsibility for their physical, spiritual and mental wellbeing. However, it was the specific conception and utility of art propagated by Ruskin, and to some extent contemporaries Matthew

Arnold and William Morris, plus their growing number of disciples, that permitted these new cultural policies to seed and become embedded in industrial towns and cities. This was the idea that paintings and artworks had transformative properties, not just as instructional models for technical skill, as was the case for ornamental design objects and their place in emerging Schools of Art and Design, but as heuristic and communicative devices that were legible and accessible to all, transmitting and creating meaning:

> This idea that art was "readable" and could thus speak to anyone, regardless of education level, relied on treating paintings as "windows" onto their subjects, making the surface of the painting itself transparent. (Woodson-Boulton 2007, p. 50)

Ruskin saw contingency between a healthy and potentially more egalitarian society and its artistic and cultural production; art offered a means of redemption from the corruptions of industrialisation and its effects on the structures and social stratification of society. Every person should, therefore, have access to art, and museums had the pivotal role of providing this access, not just for technical education but for moral edification. There were contradictions in Ruskin's views over his lifetime: he is cited in 1867 as stating that museums are not for shelter against "rain or ennui" and should not let in "the utterly squalid and ill-bred portion of the people" (in Waterfield 1994, p. 40). The poor would simply not have the arts literacy to benefit from entering museum buildings since Old Masters required technical education to be understood. In fact, his doctrine proposed a bifurcation of museums for the nation – one set for the curation and display of 'great artworks' and another for education and teaching. However, it was his insistence on the physical location of museums of education in proximity to the working populations, within places that suffered the greatest ills from industrialisation, that became the most coherent influence on new forms of arts education within the municipal museum. One important example was Ruskin's own experimental museum, the St George's Museum, in Walkley in Sheffield.

THE COTTAGE ON THE HILL

Ruskin's connection to the city of Sheffield in the north of England took the form of an evolving social experiment, which, like much of his prolific writing, brought together ideas relating to work (in the sense of manufacture), land and arts education. His interest in improving the lot of the working classes of Sheffield, whose labours were primarily tied to the city's industry of ironwork and cutlery production, and in creating what he called a National Store, for the collective conservation of material wealth and 'treasures' such as artworks and literature, coalesced with the opportunity to buy land, with personal acquaintances and with Ruskin's love of the surrounding natural landscapes of Yorkshire (Hewison 1981). This confluence of opportunity and circumstance brought into being new institutions: Ruskin founded the Guild of St George in 1871 as a fund for 'the buying and securing of land in England' through which to realise these ideas for the "liberal education of the artisan" (Barnes 1985, p. 4) pump-primed by a tenth of his own fortune and involving an accompanying group of volunteers. The Sheffield experiment also involved the purchase of St George's farm in Topley and the creation of a new museum.

The first site for Ruskin's museum in Sheffield was not in a park but was chosen because of the vertiginous view and access to clean air it offered to its intended visitor, the working man. A cottage on a hill in the Walkley suburb was bought in 1875 and became the new home of Ruskin's former student of engraving and the museum's first curator, Henry Swan, who had moved to the city (Ruskin at Walkley, n.d.). Named originally the St George's Museum, this was not meant to be an isolated enterprise but was conceived as one element in a wider programme of social reform, which aimed to entice Sheffield's metal and cutlery manufacture workers up the steep hill chosen "not to keep the collection out of smoke, but expressly to beguile the artisan out of it" (John Ruskin, Letter in The Times, March 6 1883, cited in Waithe 2013, p. 43). The museum displayed a collection of "'pretty things' for the workers to see" (Pullen 2020, p. 9): minerals, artworks, books and antiquities which were predominantly from Ruskin's own vast personal collections, with Henry Swan, and his growing family, as live-in curators (Frost 2013). The cottage building became cramped, however, and although it was expanded in 1885, further plans to build a new museum fell apart as Ruskin's mental health declined.

The Sheffield Corporation had supported the museum's mission since its inception, and as the collection grew, they explored various opportunities and sites for new premises with funds raised by public sub-scription. However, these plans were scuppered by Ruskin's persistent dis-missal of site suggestions which were close to housing and his refusal to relinquish ownership of the collection. The situation was finally resolved when the museum moved to Meersbrook Park when the hall and grounds were acquired by the Corporation in 1886. This was a strategy that had already been employed in the city's first public museum, installed in Weston House in 1875, to house the collections of a local industrialist within what is now Mappin Art Gallery, with its grounds transformed into a public park (Roodhouse 2000).

The Ruskin Museum opened in Meersbrook House in 1890 after the death of Henry Swan, ending a series of somewhat torturous negotiations over loan and storage arrangements between the two sites (Hewison 2011). The new museum was under the professional curatorship of William White, who re-organised its exhibition with a greater emphasis on Ruskin's life and work. The space and the lush green environment that the park afforded were at the expense of Ruskin's peculiar philosophies of participation, which demanded additional effort and attention from the visitor. Near a tram terminus, nestled amongst expanding working-class housing (which Ruskin had dreaded) in the middle of this impressive walled estate which boasts views over the city of Sheffield not unsimilar to those of Walkley, this new location proved extremely popular, receiving over 60,000 visitors in its opening year, with evidence of appeal to the working classes Ruskin sought to benefit (Waithe 2013). Despite its failure to stay true to the foundational principles of Ruskin's social experience, the museum remained popular until it was closed by the City Council in 1950 (Ruskin at Walkley n.d.). The collection, on repeated loan terms from the Guild of St George since its transferral, but with outdated display and dwindling relevance to contemporary urban cultural strategies, was taken into storage later by Reading University by the then Master of the Guild (Barnes 1985), and is now periodically redisplayed at Millennium Gallery in Sheffield's city centre (Fig. 3.1).

The Gallery in the Park

The story of museum-making within public parks is both strategic and pragmatic. As the sprawling city began to envelope private houses and their grounds, their purchase and occasional gifting to local authorities afforded buildings that provided shelter and storage space and a ready-made

Fig. 3.1   Meersbrook Hall. (Photo credit: Abigail Gilmore)

nearby audience for contemporaneous new museums and collections held in public trust. In Manchester, the rapid but uneven development of parks and playing fields over the nineteenth century was affected by a variety of factors that informed their design, location and facilities, combining the availability of land and legislation to convert existing properties such as cemeteries, as described in Chap. 2. Many of these green spaces were the residual estates of the wealthy, the mercantile middle class and declining aristocracy, enclosed by the growing urban sprawl of working-class housing. This built infrastructure could be transformed into other spaces for improvement and active citizenship, such as galleries and museums, addressing a "social void that could be usefully filled by civic leaders and other local notables who had pretensions to behave in ways that were publicly visible and that attempted to constitute a civic tradition that represented and contrasted with the absent nobility" (O'Reilly 2019, p. 14).

Such was the case for two of the country's first three municipal parks established by the corporations of Salford and Manchester in 1846: Peel Park in Salford, originally part of the Lark Hill estate and included its corresponding Lark Hill Mansion, and Queens Park, which had at its heart,

Hendham Hall, the home of the original landowners the Houghton family. It was also the case with the later acquisitions of land that became Wythenshawe Park, Platt Fields Park and Heaton Park, each of which contained halls that were previous seats of local aristocracy. Many of these buildings survive, and each have been or continue to be the sites of display and improvement of the kind that O'Reilly alludes to, and Ruskin welcomed, above. They also became the sites of the comprehensive set of branch museums within the city, five out of six being within public parks. Lark Hill Mansion in Peel Park became the first free public reading rooms and art gallery in 1850 as the Royal Museum and Library, the progenitor to the University of Salford which was also connected to the neighbouring Royal Technical Institute. Heaton Hall in north Manchester became used as exhibition gallery 20 years after it was acquired along with the park in 1902 by the Manchester Corporation, and Platt Hall became a local authority-owned branch museum as its management passed from the Parks Committee to the Art Galleries Committee (Manchester Art Gallery n.d.). Hendham Hall in Queens Park was demolished in 1880 to be replaced by a purpose-built gallery building which also later became a branch museum of the Manchester City Galleries.

It was not solely the opportunity presented by these empty large-windowed buildings lacking purpose that led to their use as cultural institutions for exhibition and display. As is clear from the histories of Victorian local government museum-making (Hill 2005; Waterfield 1994; Woodson-Boulton 2007, 2012), the creation of new institutions within these spaces and the collection, curation and exhibition of the objects that filled them required the intervention of policy intermediaries who could advocate for their use and access by the public. One such protagonist in Manchester was Thomas Horsfall, a fervent Ruskinian. The son of a wealthy Manchester cotton manufacturer, his ill health provided him with the freedom of commitment to his father's business and time to pursue chosen philanthropic endeavours (Eagles 2011). Years later, in 1918, he wrote to a local councillor to say that it had been specifically "formed for the purpose of giving effect to Ruskin's teaching" (Harrison 1985, p. 121). To love beauty was to reaffirm true faith, and Horsfall's commitment to a belief in the personal and wider social benefits of appreciating beauty through art also reflected his sense of personal loyalty to Ruskin.

Like fellow 'art reformers' and Ruskin followers in the regional industrialising cities, Horsfall hoped "that art museums would both represent and realise a new relationship between the government and the governed, between the middle and working classes, between beauty and the industrial city" (Woodson-Boulton 2007, p. 52). These intermediaries were brought together in Manchester as the Art Museum Committee and held prestigious connections with each other and shared pastimes of the philanthropic and cultural elite as members and representatives of the Ruskin Society, the Manchester Literary Club, the Manchester Guardian and Manchester City News, Owen's College (the forerunner to the University of Manchester) and the Manchester Academy of Fine Arts (Harrison 1985, p. 123). They included figures such J.E. Phythian, who was similarly inspired by Ruskin having visited the St. George's Museum in Sheffield, and an influential member of the Manchester City Art Gallery Committee (Woodson-Boulton 2012). Connected to other civic missions in education and improvement, they provided a comprehensive network within the shared vision of "a system which makes that kind of activity which has given them their purest pleasure" (Horsfall 1877, p. 5). This 'system', the establishment of the art museum, would bring benefits to the working classes through their encounter with middle-class virtues, teaching them to love beauty whilst simultaneously affirming the values and interests of those who inhabited (and defended) this public sphere.

A year after the St George Museum opened in Walkley, Thomas Horsfall set out his thesis on the powers of art education for moral improvement and its relationship to the natural environment in a pamphlet for an "Art-Museum for Manchester" (Horsfall 1877). The proposed scheme for this new facility was discussed by a preliminary meeting of gentleman, later to become formalised by committee, who heard of Horsfall's vision for the collections and activities the art gallery would offer and their potential impact on its hoped-for audiences (Manchester Guardian 1877). Its collections would contain paintings of natural landscapes, scenes and flora in the favoured places of Manchester's surrounding countryside, alongside reproductions of great works produced by copyists and local artists; it would specifically target the working classes of Manchester and, through art, engender knowledge of and respect for the natural world as the source "most wholesome and pure now known to our working classes" (p. 2), but which they have few occasions to visit. Horsfall noted the frustrations of the middle classes with those who "tear down branches and strew green slopes with paper and broken bottles" and who do not have the "long education" to fill them with "love and admiration of beautiful country" (p. 3). These paintings and illustrations, housed in a suitable art museum, were Horsfall's answer for those who only knew the overcrowded and smoke-filled streets of the city. This was not simply a convoluted strategy to increase litter-picking in the countryside (the access to which was increasingly restricted for Manchester's working classes), but a clarion call to the middle classes to 'do' cultural policy in terms that they could read and understand. In the pamphlet, Horsfall makes an impassioned appeal for intervention by his fellow-middle class readers, who he acknowledged have the education and the social infrastructure of the public sphere, the coffee houses and letters to the Manchester Guardian, to stand in the gap between state and civic society which addressed the market failure of private cultural provision:

> The state of our Theatres and Music Halls, our multitudinous beerhouses, show what comes of leaving national amusements to be provided according to the great law of demand and supply; the state of our towns, the foulness of their air, the demi-semi attractiveness of their parks, are results of leaving health of body and mind in them to the care of a city and its representatives. If we are not better able than the governing body of the city to guide our fellow citizens towards a higher life for heart and mind, the people which toils that we may have culture and leisure may well cry shame on us. (Horsfall 1877, p. 7)

Horsfall's proposal espouses similar approaches to music education as a form of cultural entitlement that the working classes must be trained into receiving. This would be facilitated within the art museum, through the provision of "only good music" (p. 7) already familiar to the middle classes but played repeatedly until it is ingrained in the working-class repertoire by cheaply available local musicians who would then go on to educate their successors. Horsfall sets this repertoire in sharp relief to that of the music hall in the same ways he demarcates nature and green spaces (and their mediation by the art museum contents) from the evils of the public house. The consideration of music repertoire in the moral improvement of the working classes was a matter for continued public debate in Victorian and Edwardian Manchester, from the establishment of free public concerts by Charles Halle to the predilections of the Park Committee (and especially Park Manager William Pettigrew) for bandstand concerts in the open air of Manchester parks (Gilmore and Doyle 2019). Horsfall's appeal to middle classes who read his essay and could take up his cause was carefully and ably codified in the shared familiarity with musical references that served as class boundaries, as an enticement into debates about taste and judgement.

Horsfall's Art Museum was eventually established in 1885, initially in rooms in a new purpose-built gallery building in Queens Park, in north Manchester, on the site of Hendham Hall. The protracted lead time for its installation had been complicated by the parallel negotiations between the Manchester Corporation and the Royal Manchester Institution (the RMI) and by the need for constant fundraising to create the museum's collections. The RMI had preceded the Art Museum by 65 years, established in 1820 as the foremost mechanism for Manchester to display an emerging cultural capital and taste. Raising funds through subscriptions and patronage by local and metropolitan elites, this private institution was able to build its own venue for lectures, meetings and an annual exhibition of works for sale, designed by Charles Barry, which became "a meeting place and a kind of clearinghouse for the cultural societies of Manchester" (Woodson-Boulton 2012, p. 42). Its relationship to Horsfall's mission was ambivalent, not just through its commerciality but in its mission which "remained primarily a middle-class society for appreciating the arts, rather than an active force for bringing art and culture to the uneducated" (ibid).

When the Manchester Corporation took over the RMI building on agreement to support an art collection for the city with an annual budget, this gave a green light to assuming civic responsibility of the local authority for the cultural education of its denizens, and Horsfall's case was made, allowing the agreement for the museum to move into the Queen's Park Gallery. Once in the park, however, this location was short-lived, as relationships between the Parks and Cemeteries Committee, who managed the estate, and the Art Galleries Committee broke down. According to records of meetings between the two committees in early February 1886, this was chiefly due to a clash of cultural taste and value: Harrison cites the Chairman of the Parks Committee as responding to a bust of Venus de Milo in the age-old fashion of saying his workman "could have done better" (Harrison 1985, p. 145). It was also a struggle over the opening hours of the museum, as Horsfall advocated Sunday and evening opening to provide space for the museum's intended working-class visitors to spend leisure time "happily with their children in occupations which all the members of a family can enjoy together" (Horsfall 1886, p. 14) outside of their homes and away from the pubs. Within the year, the Art Museum moved to be amid its intended audience in the slum area of Ancoats, where it found a less bespoke but enduring home for the next seven decades. Horsfall's ambitions for using art as a means to effect social change placed children and young people as the key means through which to introduce new knowledge and ways of being to the working classes. He lobbied for change in legislation to the Education Code in 1894, which permitted schoolchildren to use educational time to visit museums, art galleries, historic buildings and botanical gardens (Eagles 2011, p. 134). The Manchester Art Museum continued this focus through a children's theatre and innovative school loans scheme (Eagles 2011).

PROVINCIAL PARK MUSEUM-MAKING

Despite his sometimes ambivalent relationship with the cultural public sphere of Victorian Manchester, Horsfall's influence was also felt outside of the slums of the city, in the nearby silk town of Macclesfield. Marianne Brocklehurst, part of the Brocklehurst dynasty of silk manufacturers and benefactors of Macclesfield, was a philanthropist, collector and erstwhile adventurer. She had herself been involved in promoting children's education, frequently visiting the rural school in the nearby village of Wincle where she lived with her female companion (and presumed life partner) Mary Booth, with whom she also travelled extensively in Egypt and with whom she shares a grave in the churchyard next to the school. The two 'MBs', as they were known, were influential in British archaeology, contributing to the Egypt Exploration Fund with its founder and their fellow traveller Amelia Edwards and amassing a collection of artefacts that they brought back to Macclesfield (Serpico 2016). Marianne Brocklehurst set out lobbying the local corporation in 1994, offering to pay for a new museum (Griffiths 2006) to house the collection along with her brother, Sir Peter Pownall Brocklehurst, and the support and advice of Thomas Horsfall, who had moved to nearby Rainow, and was evidently getting involved in local matters. The site she preferred for the museum was in West Park, the first park in the town which had been founded in 1854 by public subscription and landscaped by William Barron, which also retained an exhibition pavilion featuring the architectural designs of Pugin (Historic England 2001).

There were struggles with the local corporation over the site, design and governance of the museum, which at one point saw Brocklehurst's offer of funding withdrawn, the location of the park being one of them. The West Park location was of fundamental importance for Brocklehurst, in seeing through her architectural preferences (to complement existing buildings) and for the benefit of visitors and sponsors of the museum. As she wrote in correspondence to her brother:

> I thought you and Mr Mair both agreed that a museum up a street or in a town generally contained about three persons or even less; and I am quite of the same opinion, and I really make no interest in towns museums for that very reason. As for it being "more instructive" by being situated in Park Green [not a park, but a site in the town centre adjacent to other new civic institutions, the School of Art, Useful Knowledge Society and Public Library], why the School of Art has been there for twenty years and look at the Pomp!! If you go for another building to match those on the Green you

will have to begin all over again with the Town Surveyor for architect and the old hubbub in the Macclesfield Courier to follow, and you will see me gracefully retiring once again.

What is wanted is a nice moderate sized room in The Park, a very good situation for [exhibition] of paintings and loan things from private houses and then people will turn in. And there is no reason why the Scholars of Park Green should not walk in it and get their lecture and a breath of fresh air at the same time. The School of Art and Science has already a Museum Room for South Kensington exhibits and it is always empty when I go in it. There is no need for another room there that I can see (they have a large lecture room besides). We can talk it out on Saturday with Mr Horsfall. (Letter from Marianne Brocklehurst to Peter Pownall Brocklehurst, 25 November 1896)

Brocklehurst insisted that the designs for the museum should be informed by trips to the Whitworth Gallery in Manchester and based on the designs for the South Kensington museums by their architect, Mr. Purdon Clarke. She also made exacting requirements for the position of the museum near the entrance to the park, for a veranda on its park-facing wall and for long paved paths which were "good enough for anything, especially sight-seers and trippers to walk upon". There seems to have been continual resistance, however, and an independent committee was established in 1895 to ensure the development of the museum "should follow proper lines" (Letter from Marianne Brocklehurst to Mr. Mair, 18 February 1895). Brocklehurst made this committee a condition for going ahead with the scheme, which she was sponsoring to the cost of £1000 for the building and an endowment of £100 per year, alongside the need for confidentiality, including signed affidavits that the local newspaper, the Macclesfield Courier, would not publish any details of the museum plans without prior approval.

Along with Horsfall, the initial committee included a Mr. Woodward, Mr. Fountain, Canon Bell, of Alderley, and Sir Graeme Elphenstone. It was joined by invitation from Brocklehurst by her solicitor, Mr. Mairs, who had consulted with Town Clerks in Macclesfield and in Salford (with respect to the management arrangements of Peel Park and its associated museum) and confirmed that according to the Public Libraries Act of 1892, the local corporation could delegate powers and duties to such a committee; he also advised that the proposal for its formation was made through her brother, Sir Peter Pownall Brocklehurst. Arguably motivated by both her gender and sexuality, establishing a break-wall from the local authority (and from publicity of the local press) was critical to Brocklehurst's involvement as the major sponsor of the museum, providing a confidential and trusted body of gentlemen to represent her position within Macclesfield's public sphere. The protracted disagreements over the location and design of the museum took their toll, causing Brocklehurst to write about her poor health and anxieties and to urge her brother to agree to hand over the final decisions for fitting out and running the museum to this "useful good suitable set of men" (Letter from Marianne Brocklehurst to Sir Peter Pownall, 22 June 1896). These strategies appeared successful in overcoming the barriers she faced in participating in the rational discourses of Macclesfield civic philanthropy as an equal: West Park Museum was eventually built to Brocklehurst's exacting specifications and opened in 1898 (see Fig. 3.2). However, whether due to ill-health, having suffered a recent fall, or to the anxieties detailed in her letters, Marianne Brocklehurst did not attend its opening, and died by suicide not long after (Serpico 2016).

BETWEEN THE PARK AND ITS MUSEUM

Whilst public parks and museums share some common purpose, they offer different spatial logics of practice in the ways they manage publics. They are both subject to technocratic administration and dependent on public funding models to realise public good and maintain property in public trust through the management of their resources. This involves also managing the ways in which visitors are encouraged or compelled to behave, and the uses they put to these civic spaces, their amenities and contents. As Rees Leahy (2012) discusses, the social and cultural efficacy of museums involves balancing the governance of two competing bodies: those of the artworks and objects in museum collections and the production of social bodies; the public value of social work effected through museum participation is consistently in tension with the protection and curating of works of art. As such there are strategies and conventions that manage museum participation, which, she argues, can lead to frustration, anxiety and even illness. As discussed in earlier chapters, participation and behaviour in parks is also subject to surveillance and regulation, as well as serendipity and encounter, but encompasses a breadth of constituencies brought together in everyday participation distinct from museums' conventional audience demographic. In this final section, I turn to examples of contemporary museums in parks within Manchester to explore how they navigate

Fig. 3.2 West Park museum, 1898. (Drawing by T. Roylance Lawton showing the modest design promoted by Marian Brocklehurst, Image courtesy Macclesfield Museums @The Silk Heritage Trust)

these differences, the strategies they use to bring the park into the museum and vice versa, and the potential they have for facilitating cultural and counter-public spheres, beginning with the Whitworth Art Gallery.

The creation of Whitworth Park and Institute in 1890 was led by the legatees of Sir Joseph Whitworth, who entrusted three executors to make a scheme, freely accessible and for public benefit of the people of Manchester, from his vast fortune gained through engineering design (including the standard thread for screws and the 'sharp-shooting' Whitworth Rifle). This scheme was for the Whitworth Park and Institute, which began as a gallery attached to the private home of Grove House, and was described on opening as "A worthy memorial of our late friend and neighbour, to secure at once a source of perpetual gratification to the people of Manchester and at the same time a permanent influence in the direction of technical education and of the cultivation of taste and knowledge of the fine arts of painting, sculpture and architecture" (Robert Darbishire, cited in Rusholme & Victoria Park Archive n.d.).

A Manchester solicitor, Darbishire was the leading figure in both the gallery and the park's making. He was an executor for Whitworth but also for Crompton-Potter, who owned the private estate of Rusholme House, including Potter's Field and Grove House, which sat on the edge of Rusholme outside Manchester city centre. When the land became available at Crompton-Potter's death, there was concern that this area would become more cramped housing, and Darbishire campaigned to secure the land for a pleasure ground and the site of the institute as part of Sir Whitworth's legacy. Cordial relations were established between the Whitworth Committee and Manchester Corporation's Park Committee, who advised on the transformation of the Park into multi-amenity site, with bandstand and boating lake, and access to its green spaces and planting through walkways and avenues. The costs of maintaining the park, on top of funding its redesign, put strain on the resources set aside for the Whitworth Institute, however, and in 1904, the park was handed over for management to the corporation, releasing funds back to the institute which retained five acres of parkland for future extension (Rusholme & Victoria Park Archive n.d.).

Darbyshire was aided by another executor of Whitworth's legacy, Richard Copley Christie, lawyer and former professor at Owens College (now the University of Manchester). The gallery was originally intended to be joined by a School of Art, an intention passed to the Manchester Corporation to achieve, but has from its origins been closely tied to the University of Manchester, which now owns and operates the Whitworth, and its surrounding parklands. The Whitworth's collection of 55,000 works of art, textiles and wallpaper, begun by initial endowment, are now housed in a building replacing Grove House, which opened in 1908, with a recent modern extension and renovation completed in 2015. This capital development project, funded by Heritage Lottery Fund and Arts Council England, reformed the larger exhibition spaces, increasing the environmental efficiency of the building and extending out glazed rooms and corridors presenting views overlooking the park and, to some extent, views into the gallery (see Fig. 3.3). It also included a curated 'art garden' that reached back behind the gallery into the land it had retained when the park was transferred to council ownership.

In 2015, the Whitworth appointed a Cultural Park Keeper, funded over three years through the Esmée Fairbairn Foundation, to develop and implement a new cultural park programme. The cultural park keeper focused on developing several strands of community engagement work, along with a dedicated programme for families and children, including an Outdoor Art Club and Forest Art School, several large and ambitious events like a young people's festival and an annual Frost Fair, a wellbeing programme named the Natural Cultural Health Service, and a horticultural volunteering programme. These interventions into the surrounding park space arguably increased the "contact zone" (Clifford 1997) through which to extend relationships with communities who would not normally venture into the gallery. At the time, the Whitworth was managed by Dr. Maria Balshaw, who was also the Director of the Manchester Art Gallery and Director of Culture for the City Council, a position which brought 'town and gown' together with leadership of the city's museum collections and a convening role for its cultural strategy. In 2018, a 'new broom' director was appointed when Balshaw left to take up the position of Director of the TATE gallery. A Ruskinian like Horsfall, Alistair Hudson, had previously led the Middlesbrough Institute of Modern Art, the

Fig. 3.3 The Whitworth's cafe extended into the park. (Photo credit: Abigail Gilmore)

contemporary art gallery in the Northeast of England, which like the Whitworth operated in partnership with its local university, and before that was Deputy Director of Grizedale Arts, an artist residential space on a farm near Coniston, Cumbria, close to Ruskin's Brantwood home and run on Ruskinian principles.

Building on the building renovation and programming, this change in leadership placed further attention on the role the park might play in delivering the Whitworth's mission. The cultural programmes led and resourced by the Whitworth outside of its walls, first when the building was closed for renovation and then under the direction of the Cultural Park Keeper, re-activated relationships with park users, encouraging participation in creative and growing activities and connecting with pre-existing parks groups, such as the Friends of Whitworth Park. Furthermore, the interests of the new gallery director in usership and the uses of art (as a member of the international movement, Arte Util) and his extensive knowledge and understanding of Ruskin framed the park's connections to the museum as a combination of art and nature true to its historical provenance as a "combination institution" of park and museum, which also spoke of its Manchester-ness:

> The idea of the Whitworth originally when it was founded in the 19th Century was basically as an art collection which would create visual literacy in the city and fire power the cotton industry to make better design and better products. But also to make healthy, bright, smart individuals so, yes you would go to the museum to enhance your life and use the park and enjoy nature. And through art and the study of nature and healthy living, you would become a better person. As a consequence, you would have a better society, but that then expands in terms of a city council portfolio of resources that historically includes both art galleries, a distributed art gallery system and a better parks system in the city. And the long tradition in Manchester art institutions, are of using art in order to commune with nature and both in a sort of scientific but also in a philosophical way. Basically, as the first industrial city, the idea was that one way of mediating and kind of mitigating the ills of industrialisation and pollution, was to bring the people closer to nature. (former Whitworth Director, Alistair Hudson)

Under Hudson's leadership, new ideas were introduced for more museum-led activities in the park, such as the development of a community-produced 'Ruskin Road' garden, reliving Ruskin's attempts to encourage a settlement approach employing Oxford university scholars in rural road repair in Ferry Hincksey in Oxfordshire (Cook and Wedderburn 1905). The gallery also tried to soften the threshold to the museum space and its collections, considering how best to signal the gallery entrance without putting people off entering:

> There are all these other things in the pipeline which would … in different ways, signal the intent of the institution to break down the barriers between the institution and the park. Because I guess one of the problems of having signs and kind of saying this is the gallery, this is the park … we're more interested in blurring the boundaries between those two things (Manchester Museum Partnership Research Officer, Helen Mark)

National statistics on cultural and leisure participation demonstrate significant distinctions in the rates and demographics of museum visitors to those who use public parks. At the time of the Whitworth's extension, according to national statistics, the proportion of white adults who had visited a museum or gallery *at least once in 12 months* was 53% compared with 48% of BAME adults, with an entrenched class gap of 25% between upper and lower socio-economic groups at 62% and 37%, respectively in 2016 (DCMS 2017). In comparison, parks statistics suggested over half (57%) of UK adults used local parks *once a month or more* (with 35% using it once a week) with higher levels of participation by households with children under five, and Black and minority ethnic (BAME) people (45% once a week or more, compared with 34% white residents) (Heritage Lottery Fund 2016, p. 3).

The Whitworth's participatory events and programmes in the park offered potential tools to engage with family audiences and encourage visitors who were more representative of the ethnically diverse local population to approach the museum's threshold (see Fig. 3.4). The properties of open green spaces were drawn on to encourage contact and social encounter in ways which were constrained and proscribed within the enclosed gallery space. There was evidence of success, particularly in the larger outdoor concerts and events and in visitor data which identified an increase in the similarity in ethnic profiles between gallery and park users, as well as increases in reported wellbeing from those gallery visitors who had also visited the park. Whilst it is tempting to see the park solely as a mechanism for audience development, widening participation and social practice *within* the museum, both the Whitworth's Director and Cultural Park Keeper were adamant that the park is at the heart of the museum's

Fig. 3.4   Crowds watching Circus RAJ performing at the front of the Whitworth art gallery. (Photo credit: Abigail Gilmore)

vision, as an additional site *outside* its walls in which to realise the museum's mission.

However, these assumptions were not necessarily shared. Interviews with park users[2] revealed confusion about the purpose or contents of the gallery; some were not aware that entry was free of charge or even how to get inside. There was appreciation for the museum's convenience as a shelter from bad weather; however, there was also criticism of the museum's attempts to use the park as a tool for engagement, with some preferring more conventional educational workshops in drawing and creativity, or curatorial strategies which placed artworks such as sculptures or images of paintings within the outside space of the park. There was recognition that art appreciation might provide a way to cut across the language barriers presented in social encounters in multi-ethnic neighbourhoods; however, there was also a sense that this was complicated by the intersection of the 'elitist practice' of art with the representation and inclusion of multi-ethnic linguistic communities. The boundaries of social class, ethnicity and race, language, age and life stage are visible within participation in the park, offering opportunities for mediated conviviality (Barker 2017) and cultural democracy, enhanced to some extent by the interventions and resources of the Whitworth. However, the threshold boundary of the museum was also recognised, and the inside of the gallery was perceived as an institutional space that privileged its own mission over inclusion and co-production:

> something that's more dynamic and [it] includes the community more. And also, I believe that in a way it's about demand and offer. And I feel that what the gallery is doing is what they want to offer, not necessarily what people want. (Researcher-in-residence, Ana Sanchez Santana)

Taking the Knee in Platt Fields

Platt Hall is located further away from the city centre than the Whitworth along the Rusholme road in Platt Fields Park. Once the Manchester Art Gallery costumes gallery, this early suburban branch museum was closed due to the need for repair work and fumigation of a moth infestation of the eighteenth-century hall in 2017, and took advantage of this natural pause to explore how to better use the museum space through a community engagement programme called 'Platt Hall In-Between'. This involved Manchester Art Galleries senior management, curators, collection management and engagement teams, who like the Parks team that manage the surrounding parkland and other buildings on the former estate are also employed by Manchester City Council. As one of the four remaining Manchester Art Galleries buildings, which include the Queens Park Gallery, previous home for Horsfall's Art Museum and now conservation centre, Platt Hall had been the site for display and storage of textiles, costumes and the Mary Greg collection, a diverse wealth of everyday objectives, including dolls houses and some 30,000 buttons, since it became a branch museum in 1927 (Mitchell 2018).

The closing of the museum provided the stimulus for new strategies of collection management across the puzzling jigsaw of museum spaces and collections owned by the Council. Its stakeholder communities also comprise former visitors, donors, local residents and of course park users, who were familiar with the large red brick building in the corner of Platt Fields Park and who may have visited during its varying history. The In-Between project aimed to co-produce consent about the future use of the building, negotiating with these many constituencies, whilst managing the constraints of the building and its past purpose as a costume gallery, which retains a strong attachment from previous visitors. Like the Whitworth, the closed museum has used its park as a contact zone for community engagement, working with existing and emerging voluntary groups involved in sports and recreation, nature conservation, arboriculture and horticulture, with the Friends of Platt Fields (the first properly constituted Friends group in the country) and several activist and artist groups who use the park to congregate and as a creative and inspirational space. One such alliance was formed in the park in 2020 during the periods of social distancing caused by the COVID-19 pandemic, motivated by the expansion of the Black Lives Matter movement, to gather to take the weekly kneel in solidarity, following the death of George Floyd in the United States and with respect to others murdered by institutional racism (see Fig. 3.5). The action took place on the plinth that had held a statue of Abraham Lincoln, donated to Manchester in recognition of the city's support of the abolition of the slave trade, but moved to a more central location in 1986. Like other protests and assemblies in Platt Fields (and other parks) before this, this was a powerfully visible act of locational citizenship (Di Masso 2015), instigating symbolic action, not just deliberation, as part of a counter-public sphere.

The museum responded by providing space on its windows to promote the message and motivation for the group, as part of a series of exhibitions that were co-produced with park communities in the threshold space at the front of the museum (see Fig. 3.6). This use of the windows complemented other engagement activities that broke the fourth wall of the museum, including the use of geo-locative digital games and a smartphone app to feature images of objects in the collection and interpretation linking their provenance to the history and narratives of the park.

The Whitworth and Platt Hall both use their park environments to access local communities, as spaces for serendipitous encounter and mediated conviviality, strategies with which to draw them into the museum's public sphere as constituents. As museums owned and operated by the local government, who also own and manage their surrounding parkland, they are bound into relationships with publics that are differently defined and constituted by the generative (and regulatory) practices and strategies of participation afforded within these spaces. In the case of the Whitworth, an uneasy tripartite relationship between the gallery, the city council and

Fig. 3.5 The "weekly kneel" for Black Lives Matter in Platt Fields. (Photo credit: Saira Qureshi)

the university has responsibility for holding public property in trust. The constituencies of this public sphere are stratified, following Fraser (1990), by the stakes they have in the museum and the park and the power and resources they have to affect action. As the Whitworth's Director observed:

Fig. 3.6 Platt Hall window exhibition. (Photo credit: Abigail Gilmore)

The connectivity of stakeholders is probably quite weak and ill defined. In a way it's a very kind of tentative, fragile network of constituent groups who ultimately maintain the park structure. So, the challenge in a way is to build a much more robust strategic way of working between all the stakeholders; the university, council, constituent groups, the Whitworth, to make something that really works in terms of how we have an ambition for that. (Whitworth Director, Alistair Hudson)

The Whitworth was intended as an art gallery since its inception and dominates the park created by the same industrialist legacy. It was established by a powerful set of policy entrepreneurs whose philanthropic concerns bestowed the teaching of art literacy and the organisation of pleasure grounds for the enlightenment and improvement of the Manchester people. It continues to dominate the parkland through its ownership, but also in the assumptions the museum makes about how the park serves as an extension of the museum, not for the display of artwork but rather through social practice. Its intimate ties with the university and the uneasy position of the city council as absentee landlord has left a critical gap between the local state and civil society, which is only partially filled by the Whitworth and its relationships to the park and its constituent publics, some of which are weaker than others. Conversely, Platt Fields began life as an aristocratic estate enclosed by the growing city, adopting its role as the site for a branch museum long after its public emparkment and hosting a range of separate zones and amenities that were conducive to co-production and participation. Helped by its closure, and its position tucked in the corner of this much larger multi-use park surrounded by many different coherent user groups, Platt Hall museum has been forced outside its institutional walls into contact and encounter with these different constituents.

CAN MUSEUMS BE MORE LIKE PARKS?

The observations above show the frustrations that museums face when satisfying their institutional missions for publicness. Gurian (2005) identifies the psychological concept of 'threshold fear' to help consider how this is instigated or dissolved through museum architecture and programming. For example, she notes the contradiction between museums' desire to be inclusive and their mission to provide spaces for contemplation, which often translates via architectural strategies into temple-like 'iconic' structures which are imposing and off-putting. Other physical impediments to threshold-crossing include transport infrastructure, signage and complicated entrances; however, Gurian also identifies the qualities of congregant spaces which assemble and mix both uses and users to provide neutral space for informality and serendipity, which can be planned spatially, through location and architecture and which are eased through approaches to visitor welcome and programming. These recall the open forms of Sennett (2018) and the recommendations of Gehl (2011) for encouraging life between buildings, discussed in Chap. 1. In other words, she concludes that museums should be more like public parks.

In the context of the COVID-19 pandemic, with the reverberations of lockdown and new social distancing still omnipresent, Bühler (2020) also considers this question. She points to the ingrained ambivalence over museums' relationship to the public, questioning whether the publics that are interpolated by Habermasian notion of public sphere can be fostered around museums, and how they might engender social spaces which engage rational-critical debates amongst equals. The attempts of Ruskin and his followers to facilitate critical debate about art, to provide education and moral improvement in visual literacy, can be seen as attempts to create reading publics that can or want to cross over museum thresholds. The strategies employed for creating these literacies, now understood as community engagement, audience development or socially engaged practice, as the case studies above suggest, are hampered by the museum-ness of these institutions, which creates particular and stratified publics, social bodies which are constituted not just by the improvement of the mind but by their regulation of the physical body (Rees Leahy 2012) and by the language and mode of address they employ (Bühler 2020). They are practices constrained through their relationships with the particular ways of 'doing policy', of thinking for others, which favour certain types of literacy that can be overstepped when outside of the museum walls but re-materialise on crossing the threshold.

The comparison of the relationship between the Whitworth Art Gallery and its park and Platt Hall and Platt Fields provides some clues towards the different literacies supported through forms of everyday participation within the park, and how they might be employed in support of counterpublic spheres and in democratising the spaces that museums can resource. Asked how such museums might usefully engage with their surrounding greenspace, a public art curator who had been involved in several projects within Manchester's public parks responded:

> Well, I could be blunt actually. First of all I think they should just stop themselves being cultural institutions, I think they should just be seen themselves as good neighbours [laughs] in the space, drop all their airs and graces about, you know, who they are and actually go, "Look, we all live in this space, you know, this is a space that's neglected, it's a space that could be great. What can we all do to be in this space? (Public Art Curator, Kerenza McClarnan)

Conclusion

This chapter has considered the histories and practices shared by public parks and their co-located museums in Northern England during a turbulent period of class formation and municipalisation in the mid-to-late nineteenth century, and within the recent and arguably equally turbulent contemporary context. The motivations for the curation of collections and conversion or building of these new museums share provenance with those of public parks; they also share similar advocates with social standing and entry into public spheres of their own making, although not without frequent ambivalence, dispute and dissonance. The legacy of these museum-makers continues in contemporary museums and their attempts to use the properties of the municipal park to grow and democratise their publics, in a wider context of financial and political austerity and its impact on local authority capacity and resource, as discussed in more depth in Chaps. 5 and 6.

NOTES

1. This chapter draws on research in the archives of Manchester Art Gallery and Cheshire Records Office, and follow-on research interviews with park users, local authority officers, museum and gallery directors, public art curators and researchers, supported by the School of Arts Languages and Cultures Impact Fund and the Faculty of Humanities PhD Researcher in Residence Scheme. I am extremely grateful to the funders, to all interviewees and to Luciana Lang, Ana Sanchez-Santana and Amie Kirby for their research assistance and contribution to this project.
2. Ana Sanchez-Santana undertook ethnography in Whitworth Park and qualitative interviews with park users about their perceptions of the gallery as part of her PhD placement as a researcher-in-residence in 2017.

REFERENCES

Alford, J. 2008. The limits to traditional public administration, or rescuing public value from misrepresentation. *Australian Journal of Public Administration* 67 (3): 357–366. https://doi.org/10.1111/ajpa.2008.67.issue-3.

Anon. 2011. *A walk through the public institutions of Macclesfield*, Being a series of articles … reprinted from the Macclesfield Courier and Herald. London: British Library, Historical Print Editions.

Barker, A. 2017. Mediated conviviality and the urban social order. *British Journal of Criminology* 57: 848–866.

Barnes, J. 1985. *Ruskin in Sheffield*. Sheffield: The Ruskin Gallery/Collection of the Guild of St George.

Barrett, J. 2010. *Museums and the public sphere*. Oxford: Wiley-Blackwell.

Black, A., and S. Pepper. 2012. From civic place to digital space: The design of public libraries in Britain from past to present. *Library Trends* 61 (2): 440–470. https://doi.org/10.1353/lib.2012.0042.

Bühler, M. 2020. Are museums like parks? The "public" in public museums. *Mousse Magazine*, August 22. https://www.moussemagazine.it/magazine/are-museums-like-parks-the-public-in-public-museums-melanie-buheler-2020. Accessed 10 Jan 2023.

Clifford, J. 1997. *Routes travel and translation in the late twentieth century.* London: Harvard University Press.

Cook, E.T., and A. Wedderburn, eds. 1905. *The works of John Ruskin*, Library edition, Vol. XX. London: George Allen.

DCMS. 2017. *Taking part focus on report: Engagement with museums and galleries.* London: Department for Culture, Media and Sport.

Di Masso, A. 2015. Micropolitics of public space: On the contested limits of citizenship as a locational practice. *Journal of Social and Political Psychology* 3 (2): 63–83. https://doi.org/10.5964/jspp.v3i2.322.

Eagles, S. 2011. *After Ruskin.* Oxford: Oxford University Press.

Forgan, S. 2005. Building the museum: Knowledge, conflict, and the power of place. *Isis* 96 (4): 572–585. https://doi.org/10.1086/498594.

Fraser, N. 1990. Rethinking the public sphere: A contribution to the critique of actually existing democracy. *Social Text* 25/26: 56–80. https://doi.org/10.2307/466240.

Frost, M. 2013. Curator and curatress: The swans and St George's Museum, Sheffield. Guild of St George annual lecture millennium galleries, Sheffield 16 Nov 2013. The Guild of St George, Sheffield.

Gehl, J. 2011. *Life between buildings: The uses of public space.* Washington, D.C.: Island Press.

Gilmore, A., and P. Doyle. 2019. Histories of public parks in Manchester and Salford and their role in cultural policies for everyday participation. In *Histories of cultural participation, values and governance. New directions in cultural policy research*, ed. E. Belfiore and L. Gibson, 129–152. London: Palgrave Macmillan.

Griffiths, S. 2006. The charitable work of the Macclesfield silk manufacturers, 1750–1900. PhD thesis, University of Chester, Chester.

Gurian, E.H. 2005. *Civilizing the museum: The collected writings of Elaine Heumann Gurian.* London: Taylor & Francis.

Habermas, J. 1992. *The structural transformation of the public sphere.* Cambridge: Polity Press.

Harrison, M. 1985. Art and philanthropy: T. C. Horsfall and the Manchester Art Museum. In *City, class and culture: Studies of cultural production and social policy in Victorian Manchester*, ed. Alan Kidd and Kenneth Roberts, 120–147. Manchester: Manchester University Press.

Heritage Lottery Fund. 2016. *State of the UK public parks II: Public survey, report prepared by Britain Thinks.* London: Heritage Lottery Fund.

Hewison, R., ed. 1981. *New approaches to Ruskin (Routledge revivals). Thirteen essays.* London: Routledge.

———. 2011. *Ruskin and Sheffield: The museum of the Guild of St George and its making, The Ruskin lecture 1979 revised edition.* Sheffield: The Guild of St George.

Hill, K. 2005. *Culture and class in English public museums, 1850–1914.* Aldershot: Ashgate Publishing Limited.

Historic England. 2001. Listing of West Park, Park and Gardens Grade II listing. https://historicengland.org.uk/listing/the-list/list-entry/1001495. Accessed 11 Aug 2020.

Horsfall, T. 1877. An Art-Gallery for Manchester. Manchester.

———. 1886. *The Manchester Art Museum.* Manchester: Ireland and Co.

Kershaw, A., K. Bridson, and M. Parris. 2018. The muse with a wandering eye: The influence of public value on co-production in museums. *International Journal of Cultural Policy* 26 (3): 344–364. https://doi.org/10.108 0/10286632.2018.1518980.

Manchester Art Gallery. n.d. Platt Hall in between. https://www.platthall.org/the-building.html. Accessed 06 May 2023.

McCarthy, T. 1992. Practical discourse: On the relation of morality to politics. In *Habermas and the public sphere,* ed. Craig Calhoun, 51–72. Cambridge, MA: The MIT Press.

McGuigan, J. 1992. *Cultural populism.* London: Routledge.

———. 2004. *Rethinking cultural policy.* Maidenhead: Open University Press.

Mitchell, E. 2018. 'Believe me, I remain…': The Mary Greg collection at Manchester city galleries. PhD thesis, Manchester Metropolitan University, Manchester.

Navickas, K. 2015. *Protest and the politics of space and place, 1789–1848.* Manchester: Manchester University Press.

O'Reilly, C. 2019. *The greening of the city: Urban parks and public leisure, 1840–1939.* New York: Routledge.

Pan, D. 2014. The west as rationality and representation: Reading Habermas's structural transformation of the public sphere through Schmitt's theory of the partisan. *Telos* Fall (168): 64–84.

Pullen, L. 2020. The joy of pretty things: a museum for Sheffield's workers. *Journal of Art Historiography,* Number 22 June 2020.

Rees Leahy, H. 2012. *Museum bodies: The politics and practices of visiting and viewing.* London: Routledge.

Roodhouse, S. 2000. The wheel of history—A relinquishing of city council cultural control and the freedom to manage: Sheffield galleries and museums trust. *International Journal of Arts Management* 3 (1): 78–86.

Rusholme & Victoria Park Archive. n.d. Whitworth park and gallery. https://rusholmearchive.org/whitworth-park-and-gallery. Accessed 04 May 2023.

Ruskin at Walkley. n.d. Ruskin at Walkley: Reconstructing the St. George's Museum. https://www.ruskinatwalkley.org/. Accessed 08 May 2023.

Scott, D. 2002. Music and social class in Victorian London. *Urban History, Special issue: Music and Urban History* 29 (1): 60–73. https://doi.org/10.1017/S0963926802001062.

Sennett, R. 2018. *Building and dwelling: Ethics for the city.* London: Allen Lane.

Serpico, M. 2016. *Beyond beauty: Transforming the body in ancient Egypt.* London: 2 Temple Place.

The Manchester Guardian. 1877. Article 5 — No Title, The Manchester Guardian, 13 Dec 1877, page 5.

Waithe, M. 2013. Ruskin and the idea of a museum. In *Persistent Ruskin: Studies in influence, assimilation and effect,* ed. Keith Hanley and Brian Maidment, 33–52. London: Routledge.

Waterfield, G. 1994. *Art for the people.* London: Dulwich Picture Gallery.

Woodson-Boulton, A. 2007. "Industry without art is brutality": Aesthetic ideology and social practice in Victorian art museums. *Journal of British Studies* 46 (1): 47–71. https://doi.org/10.1086/508398.

———. 2012. *Transformative beauty: Art museums in industrial Britain.* California: Stanford University Press.

CHAPTER 4

The Social Lives of Public Parks

Abstract This chapter considers the contemporary lived experience of the municipal public park, and the social values and meanings that become attached to it through the narratives and practices of everyday participation. It draws on analysis of qualitative data from household interviews, ethnographic fieldwork, stakeholder consultation and participatory workshops which took place within six ecosystem case studies in England and Scotland, as part of the *Understanding Everyday Participation* research project. Through thematic analysis of the data, the chapter considers how people articulate and navigate their relationships with public space, and with the social interactions and aesthetic, physical and symbolic experiences that parks afford. Drawing on Barker et al.'s (Int J Law Context 15(4):495–514, 2019) consideration of urban parks as 'spaces apart', which host conviviality and encounters that recognise and reproduce social boundaries, I consider the social lives of parks, and their role in practising place and public space.

Keywords Everyday participation • Mixed-methods research • Participation narratives • Mediated conviviality • Cultural ecosystems • Structures of feeling • Public space

A. Gilmore, *Culture, Participation and Policy in the Municipal Public Park*, Palgrave Studies in Cultural Participation, https://doi.org/10.1007/978-3-031-44277-3_4

103

INTRODUCTION

As discussed in Chap. 1, the significance of the municipal public park to cultural participation in everyday life became quickly apparent through the early stages of the AHRC Connected Communities research project, *Understanding Everyday Participation* (henceforth UEP project).[1] Participants in the research reflected generously and with insight on the memories and values attached to their local parks, often in direct contrast with the (less frequent) references to more recognised forms of arts and culture. In interviews, ethnography and stakeholder workshops, participants formulated connections between these participation narratives (Miles 2016) and their social interactions in the places that they live and across broader expanses of time and space. The municipal public park emerged as an important site for the practices of everyday participation, scaffolding sometimes fleeting, sometimes intimate relations which traversed communities, bonding social networks, and to some extent bridging generations and different demographics within the cultural ecosystems we were researching. The park emerged as a research site that could tell us something about its significance to its constituent users, and its own geopolitical context, revealing the governance of place and people through the meaning and values attached by their everyday participation.

This chapter focuses on the contemporary lived experience of municipal public parks, presenting empirical research findings which contribute to a critical understanding of their social value and the role they play in everyday lives of their constituents. It draws on analysis of qualitative data from household interviews, ethnographic fieldwork, stakeholder consultation and participatory workshops which took place in the six ecosystem case study sites of the UEP project, between 2012 and 2017. These sites were the two adjacent city wards of Cheetham, Manchester and Broughton, Salford in northwest England; a central area of Gateshead, on the north east coast of England, which encompassed the wards of Bensham, Saltwell, Low Fell and Chowden; a network of towns and villages on the National Park of Dartmoor; the English midlands city of Peterborough; the peri-rural village of Peterculter in the hinterland of Aberdeen, and the Western Isles of Harris and Lewis and South Uist off the coast of Scotland. Through thematic analysis of qualitative data, the chapter considers how people articulate and navigate their relationships with public space, and the social interactions and aesthetic, physical and symbolic experiences that parks afford. Drawing on Barker et al.'s (2019) consideration of urban parks as 'spaces apart' where conviviality and encounters between park users can host diversity and express difference, the chapter explores the lived experiences, memories and narratives articulated within the research to critically assess the social lives of parks.

I consider these findings within their broader place contexts, to explore how everyday participation in parks acts in relation to sense of place and identify the connections to broader policy processes within these places that 'mediate conviviality' (Barker 2017). In doing so, I take up the recommendation that "insights from research on everyday practices of living, interacting and negotiating with difference—directly or indirectly—in a variety of public spaces needs to play a more central role, informing local and national policies and practices" (Barker et al. 2019, p. 15). I also confirm the ways in which the emplaced practices of participation in public parks inform locational citizenship (Di Masso 2015) whether by abiding by the rules and conventions of park usership, or through their disruption. Participation practices in public parks demand reflexivity and negotiation, even when they are ostensibly everyday and mundane. They are generative and realise value not just for the individual participant but more broadly as a contribution to the cultural public sphere. As will be discussed further in Chap. 5, park users participate in the governance of the park as commons, contributing to collective guardianship, access and enclosure that accompanies a public asset or common-pool resource (Ostrom 1990) held in public trust, and for the vast majority of public parks in England and Scotland by local authorities.

The chapter begins by outlining the ecosystem case studies where research took place, to provide further detail to the context for participation and associated values. I then go on to discuss how research participants articulated the attachments they have with local parks and green spaces, how they value the meanings and significance of these attachments and parks contribute to their sense of place and making of public space through the participation practices they contain and engender.

CULTURAL ECOSYSTEMS AND EVERYDAY PARTICIPATION

For the UEP fieldwork, we identified six case study 'cultural ecosystems', chosen as entry points for empirical research that explores the connections between forms and practices, policies, institutions, space, place, community and economy. Mixed method research involved quantitative data mapping and analysis, durational household semi-structured interviews, focus groups and workshops and archival research in each case study site. The sites were framed as 'cultural ecosystems' to highlight the dynamic, interconnected and scalar dimensions of cultural policy flows and processes through which participation is promoted, regulated and shaped within localities. Cultural ecosystems are not necessarily bound to geography (Barker 2019); however, we were interested in the situated practices of participation, even where they may connect to creative and cultural flows and networks based elsewhere. A 'straw matrix' was used as a framework which plotted participation against investment, through which to decide the four English ecosystem case studies within specific places (see Table 4.1). The metrics for the participation axis were provided by statistical data on participation rates within the populations of places; this was the first 'straw', since these are exactly the indicators that the project railed against. The principal use of these data in contemporary cultural policy was to claim a cultural deficit in those places that were bottom of a league of participation in recognised arts and culture. The second axis of investment, or 'straw' measure, called for a more nuanced consideration of the history and narratives of investment in the arts and culture through formal cultural policy mechanisms, such as Arts Council funding, the perceived importance and centrality of arts and cultural venues and programmes within local strategic priorities, and recognition of cultural wealth by

Table 4.1   The 'Straw Matrix' of participation × investment

designation and destination status (see Table 4.1). We also wanted to take into account different topographic, demographic and geographic features within the selection of case studies, and to consider how scales and levels of governance might operate within different place types. The addition of two case studies in Scotland negotiated with the funding partner, Creative Scotland, allowed us to consider further geographies within the scope of the UEP project's budget and time constraints.

The resulting ecosystem case studies were phased over the five years of fieldwork, with further follow-on activities taking place variously in these sites after fieldwork had ended. What follows is a pen-portrait of each case study to provide context for discussion of research findings concerning parks, everyday participation and cultural value.

MANCHESTER-SALFORD

Characterised by our diagnostic framework as 'High Investment: High Participation', the "original modern" city of Manchester (Marketing Manchester 2009) and its conjoined sibling city, Salford, nested within Greater Manchester, a large conurbation in the Northwest of England with around 2.8 million inhabitants and a history of civic engagement and pride that puts culture and creativity in the forefront of the place branding and local governance (GMCA 2019). As the UEP study started, Manchester City Council had recently invested in a new cross-art form centre, HOME, alongside its longer-term support for Manchester International Festival, Manchester Art Galleries and a range of cultural venues and organisations which lead the 'Cultural Ambition' for the city, the name of its sector-led cultural strategy at the time (MCC 2010). Salford, the smaller city with a very different geography, a population around half the size of Manchester and with commensurate cultural budgets, lacks a clear centre, and lies butting up to the west of Manchester's city centre legal and retail districts, divided in part by ring roads and the Rivers Irk and Irwell. Like Manchester, during the late 1990s and 2000s, the city had also consolidated its approach to culture-led regeneration, through creative and culture-led regeneration of the previous shipping port and docks, Salford Quays, through the development of the Lowry art centre and MediaCityUK, the large cluster of media and creative enterprises centred on the move of BBC departments to the region.

The case study was the first tranche of UEP fieldwork to begin, focusing on the adjacent wards of Cheetham in North Manchester, and

Broughton in East Salford (see Fig. 4.1). Governed by different local authorities (Manchester City Council and Salford City Council respectively), we selected these areas because they provided the opportunity to hear from residents within wards close to Manchester city centre and the cultural amenities located within them, and to observe their everyday lives and practices within their inner-city neighbourhoods. Cheetham is characterised as an 'isolate' neighbourhood with high indication of social deprivation and inhabitants that are trapped into movement between similar areas of social need (MIER 2009) and one of the widest ranges of spoken languages other than English in Europe; Broughton is home to a large orthodox Jewish community. Both areas have been subject to consecutive, ongoing waves of in-migration, related to rapid industrialisation of textiles manufacture during the nineteenth century and to post-industrial re-settlement patterns, and the resulting ethno-cultural and religious diversity of their inhabitants is reflected in the large number of places of worship within the areas (Gilmore 2017). Housing stock comprises nineteenth-century terraced housing and twentieth-century residential estates,

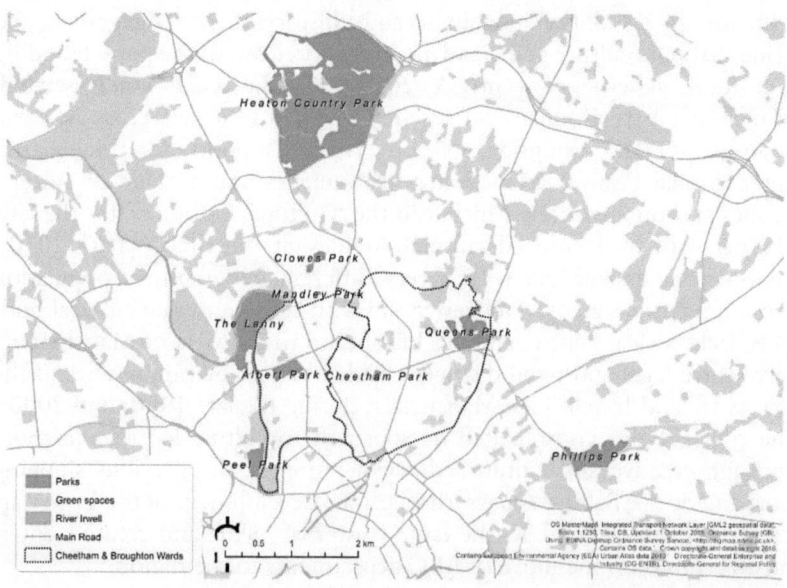

Fig. 4.1   Green spaces in Cheetham and Broughton. (Gilmore 2017)

peppered with industrial areas and brownfield sites waiting for development. There are arterial roads leading out from Manchester city centre
that traverse both areas, which are both well-served by open green spaces,
with a range of parks, nature reserves and other in-between green spaces
so far preserved from development. According to the Green Space Index
(Fields in Trust 2022), the fieldwork area scores highly in its provision of
green space, with between 50 and 70 square metres per capita within
Higher Broughton and no more than ten-minute walk for most of the
population. Indeed, the case study site is bookended by two of England's
first wholly municipal public parks, Peel Park and Queen's Park, discussed
in Chap. 2, established in the mid-nineteenth century to serve the working classes of these newly industrialised cities (Gilmore and Doyle 2019).

GATESHEAD

The second phase of fieldwork was in Gateshead in the coastal Northeast
region of England, which we characterise as 'High Investment: Low
Participation'. Another conjoined place, adjacent to the city of Newcastle,
the large town of Gateshead has historically poor provision for cultural
and leisure facilities. Since the 1980s, however, it has been the site of significant capital investment related to a long-term strategy for culture-led
regeneration programme, linked to the bidding for cultural designations,
including the European Capital of Culture, and new cultural flagship
buildings, such as the Sage Gateshead music centre (Miles 2004) as part
of a locally coherent, community-centred cultural policy (O'Brien and
Miles 2010). However, according to the predominant metrics for local
cultural participation at the time, the Active People survey placed
Gateshead within the bottom 20% nationally for its levels of cultural participation, ranked 301st out of 354 local authorities by this data in 2010
(Gibson et al. 2014). Ethnically diverse and a self-proclaimed 'working
class town' (O'Brien and Miles 2010), Gateshead has a landscape shaped
by the declining domination of heavy industry and coalmining, leaving
brownfield sites and low levels of economic activity spread over a large
urban and peri-urban space (Gibson 2019). The fieldwork focused on a
"corridor of 'situated participation' " (Gibson et al. 2014, p. 1) which
traversed the four wards of Bridges, Saltwell, Low Fell and Chowdene and
which offered proximity to 'formal' participation assets and 'major' leisure
offers: Saltwell Park, the Quayside and other civic amenities such as the
library, leisure centre, gallery and the Metrocentre shopping centre. The

fieldwork research found the lack of a vibrant town centre and the peri-urban geography of Gateshead meant that everyday participation activities happened at ward level within community settings with forays to 'town' (meaning neighbouring Newcastle) made by particular groups, such as younger people, seeking more diverse shops or the formal cultural offer of the Newcastle Quayside. In terms of green space provision, the study site in Gateshead also has good access for local residents with all being within a ten-minute walk; however, there is considerably less green space avail-able per capita, except for the ward of Saltwell which benefits from the large Saltwell Park (Fields in Trust 2022).

DARTMOOR

By contrast, the case study on Dartmoor, a National Park in the Southwest of England, focused on the rural settlements of Moretonhampstead, South Brent, Princetown, Postbridge, Heathfield, and Widecombe. These ham-lets, villages and towns form a small ring on the remote moorland, each surrounded by open green space and at some distance from the larger conurbations with shops, businesses and cultural amenities, such as Exeter, Plymouth and Totnes. The area has a demographic profile that combines both relative deprivation and affluence, with Dartmoor having a higher socio-economic profile, and an older and better-qualified population com-pared to the Southwest, and England overall, but with pockets of depriva-tion at the edge settlements (for older residents) and within village centres, for younger families with children (Milling et al. 2018). As an isolated rural region that has been neglected by both national and regional invest-ment in favour of urban cultural hubs, we classified this case study as 'Low Investment: High Participation'. Despite a lack of formal arts and cultural venues or publicly funded activities, this classification is supported by the Active People measures for arts participation. The percentage of adults per local authority who have either attended an arts event or participated in an arts activity at least three times over 12 months for leisure purposes was recorded within the districts within Dartmoor National Park as South Hams 54%, Teignbridge 49% and West Devon 48%, all comfortably above the national average of 44% (Arts Council England 2016). The types of everyday participation noted through fieldwork reflect both the remote setting with easy access to the natural heritage of its local landscape, and resistance to the dominance of metropolitan arts interests and practices within cultural policy. They include self-produced amateur drama and volunteering with professional community tours of music, dance and drama, celebrations of places and sites in the community such as fairs, carnival and spiritual performances, such as healing ceremonies in the natural landscape, as well as more common everyday practices such as pub-going and rambling. The researchers note the importance of connectivity both physically through public transport and through Wi-Fi and mobile phone coverage, particularly for young people, compounding inequalities of access to participation opportunities, and the importance of voluntary labour running projects and community spaces to sustain these activities (Milling et al. 2018). The spatiality of these assets is captured in Fig. 4.2.

PETERBOROUGH

A small city with 215,700 inhabitants in its unitary authority (ONS 2022) in a central and well-connected location in England, Peterborough is a city of contrasts. It has had over 4000 unbroken years of human occupation,

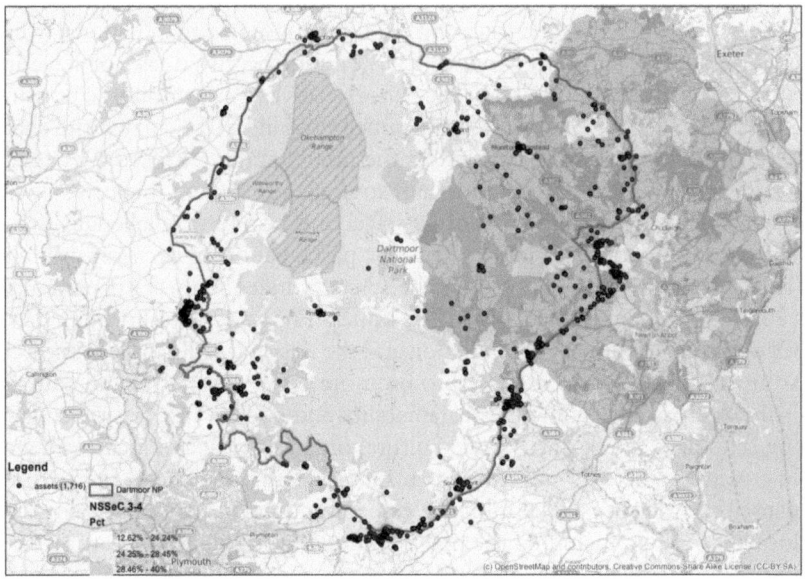

Fig. 4.2 Cultural assets mapping in the Dartmoor area, including commercial assets such as shops, cafés, post offices, pubs; leisure centres, parks, playgrounds; schools and playgroups; churches; village halls, community centres, libraries, museums, by Orian Brook. (Milling et al. 2018)

and has an historic city centre with a Norman Cathedral, ringed by new townships built in the 1960s and 1970s and surrounded by a rural hinterland with higher levels of affluence. Despite its excellent rail and road links, the city has high levels of poverty: 38% of children were living in households with below 60% of median income in 2019–2020 (End Childhood Poverty 2021). Around 18% of the population is non-white, and the city has a history of international in-migration with 20% of the population born outside of the UK (Peterborough City Council 2012a, b), including one of the largest Italian communities in the UK, who settled there to work in brick-making and construction in the twentieth century, joined by more recent European migrant workers. In 2011, 28% of primary, 21% of secondary school pupils and 16% of households in Peterborough had English as a second language (Peterborough City Council 2012b).

The city's employment specialisms include food, construction, publishing and finance. However, skills and qualifications levels are comparatively weak, with lower levels of those in professional and managerial occupations, and substantially higher proportions in elemental occupations or economically inactive due to caring for home or family than national averages (Peterborough City Council 2012c) and the city does not have its own university. Low cultural participation levels of 39%, within the bottom 20% of all places in England (Arts Council England 2016) match the challenging socio-economic statistics. At the time that the UEP project began, there were multiple regeneration and strategic investment initiatives operating in the city aiming to address social and economic challenges, fractured communities and lack of trust or participation in decision-making. The £1m Royal Society for the Arts Citizen Power programme (2010–2013), co-funded by Arts Council England, Peterborough City Council, Esme Fairbairn Foundation and the Arts and Humanities Research Council, amongst others, aimed to promote active citizenship and address issues such as environmental sustainability and addiction recovery, engaging schools networks, artists and cultural organisations, and local residents in creative gatherings and experiments using artists residencies and participatory arts practices. The initiative came at a time of fiscal constraints for local authorities, including the new coalition government's plans to reduce public spending through austerity measures and an appeal to Big Society, and included plans to green the city and foster creativity, resilience, self-sufficiency and an 'economy of regard' within Peterborough citizens

(Mclean 2010, p. 22). Whilst new models for drug recovery and stronger partnerships between creative and public bodies were realised, a scrutiny report for the City Council commented on the poor communications and lack of evidence of substantial community benefits coming from the programme, with the 'Arts and Social Change strand of the Programme [being] perhaps the most controversial at times, sometimes feeling elitist and out of touch' (Peterborough City Council 2013, p. 22).

Two Scottish case studies were chosen to reflect the interests of Creative Scotland, who co-funded the research in remote rural and peri-urban communities. These were not part of the original planning for the case studies and were not 'classified' a priori by the participation-investment matrix (although they can be understood as low investment, high participation through the interview analysis).

PETERCULTER

The village of Peterculter is in Aberdeenshire near the far east coast of Scotland. Technically within the city of Aberdeen, an oil wealthy city with a significant cultural offer, Peterculter, away from the coast and called 'Culter' by locals, was chosen because of its position on 'the edge' of the city and the Deeside oil wealth corridor. It provided an opportunity to explore the impact of the oil industry as a source of both civic identity and cultural dislocation, in the context of Culter's transition from an industrial village, with many inhabitants working in its now closed paper mill, into a commuter suburb (Miles and Ebrey 2015). The proximity of the oil industry proved initial assumptions of affluence, with statistically higher proportions of the population in professional and managerial classes than Aberdeen, but also young families with dependent children and pockets of deprivation represented through social housing. As a medium-sized village with a population of around 4400, there are few formal arts and cultural amenities, but a dependency on volunteering to maintain the village's civic assets, which during the period of fieldwork included many regular cultural and social activities, operated within its village hall, church hall, Mill Club building and leisure club. This "strong associational culture" was also joined by a sense of place identity that situated the village between urban and rural, marked by the distinction in local dialect between a toonser (town dweller) or a teuchter (country dweller) (Miles and Ebrey 2017),

THE WESTERN ISLES: EILEAN SIAR-STORNOWAY ON LEWIS AND HARRIS AND SOUTH UIST

This Outer Hebrides case study offered an opportunity to understand participation in the context of geographical isolation off the west coast of Scotland and a significant Gaelic language and cultural community. Issues affecting participation include out-migration, the sparse and ageing population, the vicinity of 'community hubs', intergenerational relations, and tensions between economic development and sustainability. Despite their remoteness, the islands are, however, perceived to be culturally vibrant with an inclusive, bottom-up culture and considerable grassroots cultural activity including historical societies, live music venues and heritage groups. Many of these explicitly aim to document and pass on the cultural traditions on the islands, which are known globally for the production of Harris Tweed, made by domestic weavers from the local wool. There is also much social enterprise activity and proactive community planning, supported by Highlands and Islands Enterprise, the development agency which funds social and economic development programmes, and by the number of community land trusts that share ownership and rights to land use.

The ecosystem case study research involved ethnographic observation, long-form household interviews, consultation and discussion groups in each site. Public parks are referred to by research participants in all six locations to lesser and greater extents, reflecting the physical and topographical environment of participants, and the conversational turns between participant and interviewer. Whilst we sampled local resident populations for interview in a way that created some form of demographic representation, the resulting qualitative data cannot represent social groups, specific locales or report evidence of causal relationships. Rather they provide personal reflections and participation narratives (Miles 2016) often over life-courses and long-term memories, habits and routines.

ORGANISED LOGICS OF PARTICIPATION

Parks may be 'spaces apart' (Barker 2017) supporting serendipitous and unplanned social interaction; however, they can only be this because of the common recognition of their intended functions. The UEP project was interested in finding out what forms of participation people undertake and found that contrary to a model of cultural deficit, even where arts and cultural participation indicators suggest 'low cultural engagement', people are incredibly busy taking part in cultural practices that they value (Taylor 2016; Miles and Gibson 2016). Exercise, recreation and sports figured in many of our research participants' park participation narratives, including cricket, basketball, golf, competitive running and cycling and football, the last particularly so in Manchester.

Participation in football, from organised teams and smaller group kick-abouts to spectating, was highly present within the Manchester interviews. The social and serendipitous opportunities of playing sports in parks were important to this interviewee, who favoured the less competitive, organised participation he had in his previous neighbourhood over the need to plan ahead in Cheetham since they had to ring ahead to organise for the floodlights to be turned on:

> And we played on a Tuesday night, just a social match, for years, and it started just—, we played a three and three match and the next week it was like a five on five, and then it grew and then for years ... we'd turn up at a park. They'd have lights on. We'd work out who was there for the night and we'd split into two teams and just play. So it was all social ... when I was playing here it got a little bit—, not competitive but too—, it wasn't as social as it was, so I stopped going. (Cheetham participant)

Facilities for organised sports were valued, not just for healthy exercise but for making and sustaining new friendships, making a particular space in the lives of in-migrants for community interaction in Cheetham Hill, a highly culturally diverse area that has easy access to several parks and recreation grounds. The visibility of active participation could also indicate the vibrancy of the area and a sense (or absence) of intergenerational community cohesion:

> I hardly ever see any elder people walking around here, hardly ever, I don't know why that is. You tend to see more youngsters playing sport in the park, you don't see middle aged men 'cause they're always busy or they're stuck inside looking at the football, watching football all the time. (Cheetham participant)

For another, playing sports provides a way over overcoming language and cultural barriers coming into the area:

> [the] first time I came here, I went to the park, I didn't know anyone but now I go and play and I know everyone now. (Cheetham participant)

Whilst the facilities of particular parks—the floodlights or the marked-out pitches—are valued it is their role as social spaces, for meeting up with friends and playing sport that is most significant. For one Cheetham resident, the routes to and distances from parks, where he might not know street or park names, shaped the cognitive geography of the area. Playing sports in parks every day in summer with groups of friends, "sometimes a lot of people like twenty or thirty, sometimes ten or fifteen", was highly valuable to him not in a measurable way, and not through attachment to a particular park, but as a resource for participation that is generative, social and non-rivalrous:

Participant: It's not, somehow it's important, but I don't, it's not important for me, but it is important 'cause that's where I spend my time.
Interviewer: Should there be more opportunities for people to get together, for instance, in the parks?
Participant: I think that like, you know, the park it's a friendly place, it's not organised, but if you go, you get more chances, you know, more to get together. (Cheetham participant)

'Getting together' is not something that necessarily happens organically. It requires specific facilitation, and for many, parks were one of the few sites where community interactions take place with a degree of spontaneity and with broad appeal, acting as a counterbalance to a sense of prevailing private development and commodification of leisure time around retail and hospitality.

The area's changing, there's more developments going on like retail, there's a lot of takeaways, there's a lot more variety of stuff going on, but the only downfall is I've not seen—, like when I was younger there was a lot of football tournaments going on in the local community, like everywhere, people used to get together, there was festivals and all the local parks have just been ignored now, they don't get maintained like what they used to … the only thing that's not available is something to bring all the people together, which is not the bars and the clubs. People that aren't drinking, more family orientated and cultural orientated. (Cheetham participant)

Alongside the more enduring relationships that are sustained through team sports, festivals, and melas, parks offer opportunities to glimpse others we live alongside in our neighbourhoods, to co-exist in 'mediated conviviality' (Barker 2017) and create legitimate, nested spaces where urban populations can co-exist. They create options for recognition, in Nancy Fraser's sense of marking status difference as a form of justice (Fraser 1998), if not actual interaction. Park users are distinguished and distinguishable by their participation practices—by their behaviour, and by visible markers of identity and ethnicity, such as skin colour and clothing. This was noted within our case study areas particularly those with highly diverse populations, such as Cheetham and Broughton, where some groups would structure their participation in parks—and hence their park narratives—around the co-presence or avoidance of other specific groups (Gilmore 2017). This can mean that parks and park lives become segregated and zoned by ethnicity and race, as park ethnographers, Setha Low, Dana Tapling and Suzanne Scheld found became increasingly the case in public space post-9/11 terrorist attack, within their longitudinal study of US urban parks (Setha et al. 2006).

The park ethnography by Delyth Edwards in Manchester and Salford also revealed diverse behaviours within common spaces, which she saw as segmented by activity and passivity, and by the motivations for participants to create 'places' within parks (see Table 4.2).

Table 4.2 Park segmentation by participant and practice

Segments	Typical users, practices and facilities
Active participants	
Creating a place for families	Particularly playgrounds
Creating a place for friends	Particularly football, other ball and team sports
Creating a place for solitude	Pigeon-feeders, creating niche spaces
Rule-breakers	Particularly dog-walkers & cyclists
Passive participants	
Strollers	Cultural Users; Nature observers
Sitters	Passing time, idling
Hidden/invisible users	Evidenced by drug use, litter, graffiti
Virtual participants	
Hobbyists and interest groups	Interest in parks and green spaces supported by the internet, e.g., friends' groups, lobbyists and advocates
Virtual creatives	Those involved in memory-making, commemoration and nostalgia; valuing parks for others; creative producers and consumers of arts and literature about parks

This taxonomy is useful in identifying firstly, the active (and passive) ways that participation is used to make public space and practice place within parks, and secondly, the common rules, both implicit and explicit, that participants share and concede through their park. It recalls the discussion in Chap. 1 of what Di Masso calls the 'emplaced practice' of locational citizenship, through everyday gestures, uses of space and 'common sense beliefs' (2015, p. 80) about good and bad behaviour. Practices of participation create spaces and shape the experiences of others and the rules about what is and is not acceptable.

As will be discussed further, in this and later chapters, participation is fundamental to the development of the commons, but it also can also reveal its limits. It is not just the fences and walls, planting and landscaping that signifies the boundaries around parks and the zones within them, but also the participation practices that take place that signal whether these spaces are exclusive, commodified and enclosed. As this participant suggests, encouraging social interaction can protect parks as public spaces and contribute to their stewardship.

Interviewer: Oh yeah, well you can—, one way to create a more social opportunity is to maintain some of the local smaller parks.
Participant: Yeah, I think you'd get a lot more people outside and I think that would create sort of—, it would create the perception of it being a safe area to go to and that sort of thing ... [But in these parks] there isn't much interaction to be honest ... everyone is in their own little world. (Cheetham participant)

WHAT IS A CITY BUT ITS PARKS?

Municipal public parks have long been acknowledged to add economic value to land and property through their proximity, from Frederick Ormsted's study of the impact of Central Park on property values in New York onwards (Crompton and Nicholls 2020). In the UK, the Office for National Statistics estimates property price rises associated with the presence of functional green space within 200 metres, through hedonic pricing method and regression analysis, to be between 0.5 and 1.4% (ONS 2018). Similarly, in the context of concerns over inequalities of place, the Bennett Institute for Public Policy estimates parks add a cumulative £78 billion to the value of UK homes, with their cooling properties to conurbations raising productivity and lowering air conditioning costs, at an estimated value of £248 million in 2017 (Manley 2021). Elsewhere in the world, values of between 8 and 10% on property values adjacent to urban parks are seen to be a reasonable starting point for estimates (Crompton and Nicholls 2020).

Whilst methodologies and absolute values vary (Crompton and Nicholls 2022) and whilst it is not the value of the park land per se but its proximate effect on nearby property value (and associated tax revenue), there is consensus that the presence of public parks and their positive effects on local economies are desirable and impactful on place. It is also not simply the private equity of urban green spaces, but their public policy utility that is measured in economic terms. As was acknowledged and advocated by their Victorian founding fathers, the environmental, health and well-being effects of parks are recognised contemporaneously through willingness to pay and natural capital assessment exercises, establishing complex formulae to identify causal returns on investment, the business case for development, or protection from it. For example, a 2017 natural capital assessment exercise in London estimated that public parks defer £950 million per year in health costs, and that individual households benefit from £900 a year from park proximity (Vivid economics 2017). Fields in Trust research calculated a total economic value for public benefits to individuals as on average £30.24 per year, with over double this amount for those from lower socio-economic groups and Black and Minority Ethnic (BAME), who are statistically less likely to have park proximity (Fields in Trust 2018). Like the original park-making periods in the nineteenth century, research on contemporary structural inequalities, such as health, economy, productivity and quality of life, reveals that these are both indicated and ameliorated by the co-presence and proximity of urban parks (Agarwala et al. 2020), raising concerns about park equity for urban planners and policy makers, as will be discussed further in Chaps. 5 and 6.

For the participants in the UEP study, the value of parks within their neighbourhoods was acknowledged with no less complexity, in nuanced and circumscriptive non-economic terms, contextualised by their lived experience of place and the participation narratives of their everyday lives. Interviews frequently referred to the importance of nearby parks and critically reflected on their locations, their amenities and other qualities. Public parks were cited as aspects of what makes a 'good place' and as a source of local pride.

[A]nd the park, that new park is lovely. There's a new park being done, which is gorgeous ... down in Johnson Gardens where the council estate is, and it's a really lovely spot for a park.(Peterculter participant)

The specific properties and functions of public parks were viewed as different to, and sometimes better than, other key components of urban infrastructure, as with the participant above who observed parks provide accessible spaces where people can gather, not prescribed by drinking culture. Parks are part of participants' social imaginary of their places, contrasting with more commodified leisure sites where people gather, such as shops, restaurants, cafes and bars, and valued for their calm:

I do quite like Bretton, I think there's a lot here, there's a few schools, there's, you know, the shopping centre, there's a lot of parks ... I mean they've got—, there's more than just one park 'cause they've got the normal park, the water park and then they've also got a park there which is kind of like Bretton Park, it's part of Bretton ... Nene Park, it's a big park, very big park. There's water, you can go feed your ducks if you want and stuff [laughs]. It is just a relaxing environment and basically I like calm places, not too busy. That's not to say I am shunning the places that get busy. Peterborough city centre sometimes you want to go out, you know, to people, with somebody else and enjoy coffee. (Peterborough participant)

Peterborough has the distinction of its new town origins, resembling less a city than a collection of settlements linked together by routes through to open green spaces, country parks and open countryside.[2] This topography is manifest in the park narratives of participants, who identify the proximity to the countryside, or 'natural heritage', according to park management lexicology (City of Peterborough 2019), as a particular aspect they like about their area:

Within five or ten minutes' walk I can be in Bretton Park, or if I like to go that way I can be out in the countryside. There's some woods around here I can walk round, which I do. (Peterborough participant)

The development of the 'Mark Three New Town', designated in 1967 (TCPA n.d.), combined Ebenezer Howard's garden city and social city ideals with post-war planning for London overspill, leaving Peterborough at the end of a discontinuous urban corridor stretching northwards from the capital (Hall 1996). The conurbation has continued to expand, adding a large central shopping mall near its ancient city centre, which is ringed by the individual townships of Werrington, Orton and Bretton and large suburb of Hampton, each with their own shopping centres, linked by orbital dual carriageways and enclosing the open green space of Peterborough's rural hinterland (Grant 2017). Writing about new towns in the 1980s, Hall (1996) described new towns as a planning ideal that had been implemented wrongly—top down, by the post-war-managed welfare capitalism government, to house a growing labour force for new high-tech industries. The legacy of Howard's garden city could be observed in the location of new towns off newly built motorways in 'a serene green world … the vegetation long ago lushly enveloped them' (p. 134). Despite the dominance of the road network, which circumscribes the city centre, connecting up the townships, walking around the green infrastructure of this peri-urban landscape of Peterborough features in many interview narratives. Participants describe their routes between smaller parks and link spaces within their housing estates to the larger country parks the city hosts.

> So, erm, doing the walk round the park I would do that generally on my own, And then it became part of a—, a fitness regime in as much as that was the primary reason for doing the actual walking through the park. And it's a very nice place to go, you know … in fact when we were first married we lived here … so I've been in this vicinity all my married life, so 40 odd years, so we've seen the park from when it was brand new. It was the whole—, this whole area was new then. And so from little saplings to big trees that the kids swing on. So in terms of an environment, we've watched that change and so it's always pleasant to do that. It is—, I mean it's different now in terms of—, obviously because things have grown up there's some areas that are a little bit overgrown maybe and perhaps would feel a little bit walking through on your own, but I'm not really a fearful person like that, I don't—, it doesn't tend to […] to put me off. (Peterborough participant)

WALKING AND BOUNDARY-MAKING

The narratives on walking within green spaces reveal how participants actively use walking practices for social imagination and practising of place, even when its motivation was purely functional, for example, to take exercise or to get to a specific destination. Walking allows people to see changes in the surrounding environment, both seasonal and man-made in their personal real-time, helping them structure the passing of time. Walking gives temporal space for noticing time passing, but also for reflective thinking and for eudemonic work that may be unrelated to the immediate environment. It is a physical activity regularly prescribed for medical reasons, not only for clinical health benefits (such as losing weight or providing suitable cardio-vascular exercise) but also for the hedonic affective benefits which address mental health issues and contribute to general well-being (Arsenijevic and Groot 2017).

> I like [Bretton Park] because it gives me an opportunity to come out and connect a little bit with nature. You can even think better when you're not [inaudible 00:19:45] with the—, [mumbles] your head and you know you come out and you just take a walk and you hear birds tweeting. It sounds a bit corny and stuff but it—, I—, over my work I've seen a lot of different people that have to die, yeah. And because I'm an interpreter and you have to go to hospitals and then—, and the people that have to go to prison and the people have to—, yes. And that makes me appreciate things, you know. And when you lose, you lose it, so you better—, [Both talking at once] And it's now. It's all about now, not tomorrow, before or yesterday. It's now. It's a—, yeah, you know. So I like—, 'cause it gives me this opportunity to come out in the nature. (Peterborough participant)

These reflections are quite different to the assumptions within Victorian park-making about the instrumental benefits of public walks for the working classes: even when, as with the interviewee above, they are still related to relief from the strictures of labour, there is a sense of reflexivity and awareness of walking as a social practice rather than prescription. This fits with evidence that encouragement for walking for public health is more successful when hedonic and eudemonic feelings are emphasised as people are more likely to consistently decide to walk if they feel good about prior decisions to walk, and if they are led towards 'what feels good' rather than what is 'good for them' (Segar and Richardson 2014).

Walking practices were cast in new light within different contexts, however. In the rural, remote Western Isles, an abundance of natural heritage surrounds the small settlements on the islands. Bounded green spaces and managed parks feature far less as central community assets or places for assembly than in the urban case studies; rather, traditional cultural practices such as cèilidhs and festivals are cited as important infrastructure to bring communities together, and indoor places are treasured as sites to shelter from the often wild and treacherous weather. Participants talked about their sense of place when navigating cultural boundaries and their different positions of authenticity and status. Their experiences of belonging were fractured by anxieties about what kinds of community interaction were appropriate and whether they were welcomed, depending on whether they were 'incomers' (even of long standing) or indigenous to the islands.

One of the few bounded green spaces mentioned by island interviewees is the grounds of the Lews Castle, in Stornaway, a Victorian Tudoresque fort built by Lord Matheson who owned the island (and who made his fortune from the Chinese Opium trade). The castle was bought by Lord Leverhulme and later gifted by him to the parish in 1923, and established as a tourist destination, run by the Stornoway Trust, the first of growing numbers of community land trusts on the islands. The grounds are a significant place for town dwellers, for walking, engaging with the landscaped gardens and for the views they offer over the bay, as a secluded retreat where

> you're not going to get overheard, you're not going to be, like, tripped over by somebody on their way to do something else. (Stornaway participant)

The other managed spaces are play parks and sports recreation grounds. These are sites of struggles over decision-making and fundraising, revealing tensions between the local council, community land trusts and other interests over community land management, activism, economic sustainability and stewardship, discussed further in Chap. 5.

The narratives of walking practices in the Western Isles reveal anxieties around social boundaries: where and when walking happens can signal authenticity in relation to insider-outsider status, and in contrast to the expected behaviour of tourists. For 'locals' walking is predominantly functional not social, to take the dog out or as part of work on the land.

> Walking is not something that local people do. They don't go out for walks. Some people do go walk—, out for a walk with the dog, you meet some people out. Very few people do that. But the idea of going off with other people for a walk where you go up a hill and you read maps, God, no, you would never see anybody doing that here. That is an important part of society in Britain. But it doesn't happen here at all. (Stornoway participant)

Whilst there were expressions of the simple hedonic pleasures of walking, particularly on the beach, and particularly solo, there was also the sense that walking as an appropriate pastime was somehow risky or rebellious:

So being retired, I usually walk half an hour a day, briskly, maybe to the top of the hill. Sometimes I go to Eriskay, park the car and time my walk. But I can do that and I like doing that. I'm comfortable in my own skin so I do what I want to do. (South Uist participant)

One participant talked about the distinctive spaces for walking for work and for leisure, distinguishing between watching over farm animals and exploring the wilds, and complaining that "It's quite difficult now to find someone who wants to keep an eye on your cattle and your sheep." He went on to say:

I would like to explore much more, the Western Isles, yes, something, if one day I can find someone who would like to come along with me to explore the wild places of the Western Island, so inside of South Uist. I know that thing 'cause my wife doesn't want that I go on my own 'cause walking on the moorlands even if you don't really know the place you can—, anything can happen anyway, 'cause even if I don't go that far it's quite easy to fall or you meet a stag on the—, and the stag can go straightaway towards you. Or you know, anything can happen. (South Uist participant)

These anxieties contrast walking in wild, risky spaces, with managed land. However, the divisions between what makes private and public, protected or uncontrolled on the Western Isles is complicated by its retention of common land through the continuation of crofting and the communal ownership and management of land through land trusts:

So it's the thing where you start off with a garden that is entirely, well, [laughs] no, there is, well, obviously there's the wind and the rats and the hedgehogs come in and whatever. But it's very much under our control, isn't it? And then we walk up that lane and as you go further and further up the lane it becomes less and less our space. And you go through the gates and it's the common grazing and you've got a similar experience except that there's none of our sheep. It's just us and Tilly [dog], you know. And it's lovely. We could be picking things, you know, and so picking blackberries or gathering bog myrtle or heather, or just a walk, yeah. (South Uist participant)

The participation narratives of those on Dartmoor observed how the relationships between parks and green spaces and their sense of place are shaped by the governance framework for their National Park. There were some references to public parks within the towns where fieldwork took place, most notably Victoria Park in the small town of Buckfastleigh, which has a skatepark and 130-year-old open air pool, run by a community charitable trust who are active in local fundraising. Similar to the Western Isles, the discussion about planning and governance of green spaces identified tensions between local parish and district authorities, the county council and the National Park authority (as discussed further in Chap. 5). Unlike the Western Isles, walking was identified as common practice with much enthusiasm and participation in walking groups. Dartmoor itself was identified within these narratives as a singular natural domain that dominates the individual towns and villages.

> But I would say Dartmoor as a place is more like—, it's a bit like the difference between a forest and a city, a city is culture and a forest is nature and Dartmoor is very nature-orientated. There isn't so much a culture here, it's just a really nice place to be and it's quite a healing place in some respects. (Dartmoor participant)

The dominance of a National Park on the doorstep brought the notion of what makes a park and where it starts and ends into question:

> You just walk down and it's along the riverbank, big open area. It sort of doesn't feel like a park, do you know what I mean? It's kind of I suppose basically a park but it doesn't have that kind of parky vibe, it's not like paths and wild countries and that kind of thing. There's ponies wandering about and you know, there's the River Dart and you can go—, there's a hidey hole place, you can go down one end where there's a sort of bit where the river kind of turns into more of a big pool. (Dartmoor participant)

Across the case study sites, there was evidence that access to natural heritage can negate the need for parks and formalised green space:

> We've got lots of wild places round Culter so we don't need parks and I think—, maybe—, you know in the terms of sort of structured play parks, I don't know, my boys used to go and climb trees. (Peterculter participant)

However, those with prior metropolitan experiences placed the functional value of parks quite specifically in relation to leisure time, historically and contemporaneously.

[T]here are many beautiful walks, rivers, which I do love. But unless one has a vehicle, forget it, there are no actual just walks as you would—, I was brought up in London. We had parks, we had recreation grounds, we had edges of golf courses where you could walk to even though you're surrounded by city. But not here because if you think back mill workers wouldn't have time for leisurely walks, they were too bloody tired. (Dartmoor participant)

For others, bounded green spaces are more valued, and safer, than the 'wild' spaces in between:

this is where that area I think is going to be regenerated, where I said that the land has been empty for quite some time, and it's nice to walk through to get to Bretton Park but it is a little bit overgrown and a little bit—, so it's not somewhere, you know, I'd walk through in the daylight with Malc, but I wouldn't go there ordinarily on my own, but then there's no real need either. (Peterborough participant)

WAYMARKERS AND LANDMARKS

Parks appeared in all the case studies as local landmarks that pervade through time, evoked nostalgically by childhood memories:

I used to like going to Broughton Park, Broughton Park, Higher Broughton. But it's little landmarks where, you know, apart from that I can't—, you'll go and it'll come to me, why didn't I mention that? I should have mentioned that. No, I can't think of anything really apart from the little parks, you know where you used to go as a kid? (Broughton participant)

They held memories, of both mundane and significant life events and provided a form of constancy even when some aspects of the park had changed:

Because it has not loads of memories but it has certain memories for me ... years and years, I got caught by my mum wagging school there with a fella, a lad I shouldn't have been with, I remember when our Antony died making daisy chains with our Natalie, sat under a tree, giant tree and the tree is still there ... I remember taking the kids on the swings for them to be out of the way 'cause something had happened, but I just find it peaceful and you know why 'cause I remember the museum being there as well. (Cheetham participant)

As memory waymarkers, parks embody feelings of security, belonging and familiarity. A mother in Cheetham talked about visiting parks outside of the local area for the facilities they have for her children, but also for the memories and conversations that they elicit for her and her family. Day trips to Heaton Park, the large park bought by Manchester City Council to the north of the city boundaries in 1901 (O'Reilly 2019), are important because they connected her to her memories, allowing participation in another place from her past, endowing Heaton Park with further value:

> We are going there because there are very open areas, same like as in Indian fields … There we are playing and seeing the animals … It's a good place because when we are going there we remember India. (Cheetham participant)

For one Gateshead participant, green spaces were pivotal and formative environments throughout their life. They were born and grew up in the park-keepers' lodge in Saltwell Park, due to their father's job in park management with the local authority. A keen, life-long golfer, paying twice a week as part of a club, they were influenced in taking up the sport, having been part of the bowling club there with their father who had a job in the parks department. Interviewed outside in the garden of the house they bought when they returned to the area because of its proximity to the park, they talked about their involvement in sports clubs, and their love for entertaining friends and family outside. Playing golf was a conduit to spending time outdoors and visiting different green spaces, including those on competitive circuits, to make friends and be part of a wider professional network.

For some, parks offered the chance for volunteering, creating opportunities for social interaction and to connect with others. A Cheetham participant became active in lobbying for her local park after moving to the area, to overcome isolation. Researching its heritage and finding ways to get others involved in practising public space. She talked of plans to re-enact performances and activities that evoked participation in the park a century ago, as a means to activate further participation from others in the neighbourhood, and provide a mechanism to integrate a new community support police officer (PCSO).

> Because hundreds of years ago in Cheetham Hill, they used to have little shows every month, where they'd have brought different people in and I

suppose different gospel groups and just everyday groups, and everybody got together. Well, we're trying to get that back up and running again. But one of the bandstands that we need to use, it's a listed bandstand, so we have to go through the right channels for it before—, and we're hoping that the likes of B&Q and so on, with the plumbing shops in the area, they're going to help us out with all of the paint and whatever we need. And then we're going to get some of the groups that we're involved with to come along, like the kids from the groups, they're going to come along and do things on it. What we've suggested was that we all dress up as well in Victorian-style clothes and get the police band involved. And we have a new PCSO on the area at the minute, Craig, so we're giving him the task of—, well, we're giving him this, more or less. We've thought of it but we're giving it to him so as he can keep it, like, ongoing within the community. And it's getting him known to the people in the community as well.

These acts of civic participation in public space, or emplaced, locational citizenship following Di Masso (2015), of taking part, coming towards, stepping up and joining in, were both personally and communally transformative. This particular park, Cheetham (or Elizabeth Street) Park, is nestled within the post-industrial landscape of North Manchester amidst former textile workshops and crumbling civic infrastructure adjacent to a spinal route from the city centre up to Cheetham Hill, where the Victorian middle classes had built their villas to escape the pollution of the inner city but long since left. The listed bandstand mentioned above and another contemporaneous shelter were symbolic landmarks within the stewardship of public space in the area: they were recognised as of heritage value, but also increasingly as evidence of the lack of care for the park, both on the part of the community (since the bandstand was vandalised) and of the local authority who could not afford their expensive repair to the standards required by heritage listing (Gilmore 2017). The bandstand and shelter became focal points for successive attempts to develop community engagement which would lead to permanent social infrastructure, such as a friends' group who could care for it, lobby for further investment and protect it from potential development. Even when these attempts were not successful (Gilmore and Lang 2020), they were significant in revealing the (often contradictory) perceptions of whose responsibility it was to maintain public space, discussed further in Chap. 5.

CONCLUSION

Everyday participation in public parks held a wealth of significance for our research participants. Active participation in the urban inner-city contexts of Manchester, Salford and Gateshead provided the means for social encounter and a measure of vibrancy and cohesion of their local areas. Access to parks and green spaces presented routes around their neighbourhoods, and waymarkers for memories. Whilst most participants relayed positive experiences and meanings attached to their participation, there was also acknowledgement that public spaces can be enclosed and exclusive at certain times and for certain groups. Participants avoided spaces where they felt unsafe or when they knew that certain groups might be present, such as young people, or when they may be party to behaviours that transgress religious or ethnic conventions, such as drinking or lack of body covering when sunbathing. Small gestures and routine practices can also be misattributed: for example, the pigeon-feeding in parks by Asian families, a form of recycling and religious respect for natural life cycles, was seen as anti-social and expensive for local authorities due to the cost of waste management and vermin control. Instead of a sign of everyday participation, this is seen as active disregard for and a lack of ownership of open spaces in the neighbourhood (Gilmore 2017).

For those in rural and peri-rural contexts, the managed space of the park takes on different meanings, although even outside of park gates on the moors or the commons and crofts of the Western Isles, there were social boundaries and conventions guiding the 'proper' use of public space, many of which were internalised within the narratives of the research participants. Walking in green open spaces could be risky, rebellious, a sign that you are an 'incomer' or 'outsider', a rule breaker or rule maker, providing through your presence a comfort to others that spaces are public and welcoming, of taking care and protecting common usership, or posing a potential threat and enclosure. These distinctions are shaped by the different motivations for using and being in parks, which are in turn shaped by the different contexts of place and policy of the six ecosystems in which we undertook fieldwork. These places are characterised variously by topography and geography, by the existence of available natural heritage and managed park space, and the histories of the authorities and interests that own and regulate these spaces, the forms of work and industry of these settlements, and by the cultural traditions and identities of those that inhabit them. They are 'thrown together' places (Massey 2005) of diverse ethnicity and affluence, class and health inequalities, impacted by the politics of the 2010s, with declining social welfare, the retraction of the state, and the divisions of post-referendum Britain. These contexts in turn shape participation as what Gearey et al. (2020) identify as 'liquid leisure' after Blackshaw (2010) when they discuss the ludic qualities of participation in the green open spaces of wetlands, meaning the practices of leisure that are not solely defined by their intrinsic purpose but should be understood more holistically as expressions of the practising of self, family and, I would argue, other institutional forms within society. Through these acts of liquid leisure, of everyday participation, the participants we spoke with were practising citizenship; whether through the organised logics of team sports and park runs or simply cutting through their local park on the way to somewhere else, they were contributing to the social lives of parks and mediating these important public spaces.

NOTES

1. The project was funded by the Arts and Humanities Research Council Connected Communities 'Communities, Culture and Creative Economies' programme (AH/J005401/1) with partnership funding from Creative Scotland. As Co-Investigator on the project, I led the case study of Greater Manchester. For further information see https://www.everydayparticipation.org.

2. A 2019 assessment report finds Peterborough has above average amount of publicly available open green space, with much being natural heritage, but a shortage of medium- and large-sized parks suitable to accommodate culture and recreation amenities that would be needed as the city population and housing density grows (City of Peterborough 2019). The report also recommended specific attention be weighted towards neighbourhood parks for rejuvenation in order to support better park equity according to its measures.

REFERENCES

Agarwala, M., Y. Cinamon Nair, M.C. Cordonier Segger, D. Coyle, M. Felici, B. Goodair, R. Leam, S. Lu, A. Manley, J. Wdowin, and D. Zenghelis. 2020. *Building forward: Investing in a resilient recovery*, Wealth economy report to LetterOne. Cambridge: Bennett Institute for Public Policy, University of Cambridge.

Arsenijevic, J., and W. Groot. 2017. Physical activity on prescription schemes (PARS): Do programme characteristics influence effectiveness? Results of a systematic review and meta-analyses. *BMJ Open* 7 (2): 012–156. https://doi.org/10.1136/bmjopen-2016-012156.

Arts Council England. 2016. Active People Survey data 2009 & 2010, published 22 Feb 2016. https://www.artscouncil.org.uk/research-and-data/active-lives-survey#t-in-page-nav-4. Accessed 16 May 2023.

Barker, A. 2017. Mediated conviviality and the urban social order: Reframing the regulation of public space. *British Journal of Criminology* 57 (4): 848–866.

Barker, V. 2019. The democratic development potential of a cultural ecosystem approach. *Journal of Law, Social Justice and Global Development* 24: 86–98.

Barker, A., A. Crawford, N. Booth, and D. Churchill. 2019. Everyday encounters with difference in urban parks: Forging 'openness to otherness' in segmenting cities. *International Journal of Law in Context* 15 (4): 495–514. https://doi.org/10.1017/S1744552319000387.

Blackshaw, T. 2010. *Leisure*. Abingdon: Routledge.

City of Peterborough, 2019. Assessment of parks and opens spaces. Appendix C.

Crompton, J., and S. Nicholls. 2020. Impact on property values of distance to parks and open spaces: An update of U.S. studies in the new millennium. *Journal of Leisure Research* 51 (2): 127–146. https://doi.org/10.1080/00222216.2019.1637704.

———. 2022. The impact of park views on property values. *Leisure Sciences* 44 (8): 1099–1111.

Di Masso, A. 2015. Micropolitics of public space: On the contested limits of citizenship as a locational practice. *Journal of Social and Political Psychology* 3 (2): 63–83. https://doi.org/10.5964/jspp.v3i2.322.

End Childhood Poverty. 2021. Child poverty in your area 2014/15–2019/20. https://endchildpoverty.org.uk/local-child-poverty-data-2014-15-2019-20/. Accessed 10 Apr 2023.

Fields in Trust. 2018. *Revaluing parks and green spaces, measuring their economic and wellbeing value to individuals*. London: Fields in Trust.

———. 2022. Green space index. https://experience.arcgis.com/. Accessed 28 Apr 2023.

Fraser, N. 1998. *Social justice in the age of identity politics: Redistribution, recognition, participation*, Discussion paper FS I 98–108. Wissenschaftszentrum Berlin für Sozialforschung.

Gearey, M., A. Church, and N. Ravenscroft. 2020. *English wetlands: Spaces of nature, culture, imagination*. London: Palgrave Macmillan.

Gibson, L. 2019. Cultural ecologies: Policy, participation and practices. In *Histories of cultural participation, values and governance. New directions in cultural policy research*, ed. E. Belfiore and L. Gibson, 153–182. London: Palgrave Macmillan.

Gibson, L., M. Taylor, and D. Edwards. 2014. Understanding everyday participation in Gateshead briefing document, July 2014, published by the AHRC

Understanding Everyday Participation project. www.everydayparticipation. org. Accessed 12 Jul 2022.

Gilmore, A. 2017. The park and the commons: Vernacular spaces for everyday participation and cultural value. *Cultural Trends.* https://doi.org/10.108 0/09548963.2017.1274358.

Gilmore, A., and P. Doyle. 2019. Histories of public parks in Manchester and Salford and their role in cultural policies for everyday participation. In *Histories of cultural participation, values and governance. New directions in cultural policy research,* ed. Eleonora Belfiore and Lisanne Gibson, 129–152. London: Palgrave Macmillan.

Gilmore, A., and L. Lang. 2020. Talking, walking and making in Cheetham Park: Reflecting on everyday participation as a method and the failure of an interdisciplinary commons. *Conjunctions* 7 (2). https://doi.org/10.7146/tjcp. v7i2.11925.

GMCA. 2019. Grown in Manchester, Known around the World, Greater Manchester Combined Authority Cultural Strategy 2019–2024. https://www. greatermanchester-ca.gov.uk/media/1980/strategy.pdf Accessed 21 Dec 2023.

Grant, A. 2017. Peterborough: How an ancient city became a new town, September 7. https://alexgrant.me/2017/09/07/peterborough-how-an-ancient-city-became-a-new-town/. Accessed 13 Jun 2022.

Hall, P. 1996. *Cities of tomorrow.* Oxford: Blackwell.

Manchester City Council. 2010. *Reframing Manchester's cultural strategy.* Manchester: Manchester City Council.

Manley, A. 2021. The value of public parks, September 21. Cambridge: Bennett Institute for Public Policy. https://www.bennettinstitute.cam.ac.uk/blog/ value-public-parks/. Accessed 04 Jul 2022.

Marketing Manchester. 2009. Original modern, first published February 2009. http://www.marketingmanchester.com/wp-content/uploads/2017/02/ OriginalModern.pdf. Accessed 28 Apr 2023.

Massey, D. 2005. *For space.* New York: SAGE.

Mclean, S. 2010. *Citizen power in Peterborough: A scoping study.* London: RSA.

MIER. 2009. *Manchester Independent Economic Review: Sustainable Communities, Greater Manchester Combined Authority.* https://www.greatermanchester-ca. gov.uk/what-we-do/economy/greater-manchester-independent-prosperity-review/manchester-independent-economic-review-archive-2009/ Accessed 21 Dec 2023.

Miles, S. 2004. Newcastle Gateshead Quayside: Cultural investment and identities of resistance. *Capital and Class* 84: 183–190.

Miles, A. 2016. Telling tales of participation: Exploring the interplay of time and territory in cultural boundary work using participation narratives. *Cultural Trends* 25 (3): 182–193. https://doi.org/10.1080/09548963.2016.1204046.

Miles, A., and J. Ebrey. 2015. Understanding everyday participation: Aberdeen briefing document. https://www.everydayparticipation.org/wp-content/uploads/2017/04/UEP-Aberdeen-interim-briefing-report.pdf. Accessed 06 May 2023.

———. 2017. The village in the city: Participation and cultural value on the urban periphery. *Cultural Trends* 26 (1): 58–69. https://doi.org/10.1080/0954896 3.2017.127436.

Miles, A., and L. Gibson. 2016. Everyday participation and cultural value. *Cultural Trends* 25 (3): 151–157.

Milling, J., K. Schaefer, and D. Edwards. 2018. Understanding everyday participation: Dartmoor report, March 2018, AHRC Understanding Everyday Participation. www.everydayparticipation.org. Accessed 05 Apr 2022.

Nene Park Trust. n.d. Nene Park: Plan your visit. https://www.nenepark.org.uk/pages/category/plan-your-visit. Accessed 10 Apr 2023.

O'Brien, D., and S. Miles. 2010. Cultural policy as rhetoric and reality: A comparative analysis of policy making in the peripheral north of England. *Cultural Trends* 19 (1–2): 3–13. https://doi.org/10.1080/09548961003695940.

O'Reilly, C.A. 2019. *The greening of the city: Urban parks and public leisure, 1840–1939.* Vol. 73. London: Routledge.

ONS. 2018. *Estimating the impact urban green space has on property price.* Release 26 July 2018, Office for National Statistics https://www.ons.gov.uk/economy/nationalaccounts/uksectoraccounts/compendium/economicreview/july2018/estimatingtheimpacturbangreenspacehasonpropertyprice. Accessed 21 Dec 2023.

———. 2022. How the population changed where you live. Office for National Statistics. https://www.ons.gov.uk/peoplepopulationandcommunity/populationandmigration/populationestimates/articles/howthepopulationchangedwhereyoulivecensus2021/2022-06-28. Accessed 03 Apr 2023.

Ostrom, E. 1990. *Governing the commons: The evolution of institutions for collective action.* New York: Cambridge University Press.

Peterborough City Council. 2012a. 2011 census profile for Peterborough Unitary Authority. https://www.peterborough.gov.uk/asset-library/imported-assets/AboutPeterborough-Census-PboroUnitaryAuthority.pdf. Accessed 09 Jan 2023.

———. 2012b. 2011 census: Ethnicity, identity, language and religion for Peterborough Unitary Authority. https://www.peterborough.gov.uk/asset-library/imported-assets/AboutPeterborough-Census-EthncicityIdentityLanguageReligionKeyStat.pdf. Accessed 09 Jan 2023.

———. 2012c. 2011 census headlines economy and labour market Peterborough Unitary Authority. https://www.peterborough.gov.uk/asset-library/imported-assets/AboutPeterborough-Census-EconomyAndLabourMarket.pdf. Accessed 09 Jan 2023.

———. 2013. *Citizen power programme: Final report of the strong and supportive communities scrutiny committee task and finish group.* Peterborough: Peterborough City Council. https://democracy.peterborough.gov.uk/mgAi. aspx?ID=7830/. Accessed 09 Jan 2023.

Segar, M.L., and C.R. Richardson. 2014. Prescribing pleasure and meaning: Cultivating walking motivation and maintenance. *American Journal of Preventive Medicine* 47 (6): 838–841. https://doi.org/10.1016/j. amepre.2014.07.001.

Setha, L., D. Taplin, and S. Scheld. 2006. *Rethinking urban parks. Public space and cultural diversity.* Austin: University of Texas Press.

Taylor, M. 2016. Nonparticipation or different styles of participation? Alternative interpretations from Taking Part. *Cultural Trends* 25 (3): 169–181.

TCPA. n.d. Peterborough, Town and Country Planning Association. https://tcpa.org.uk/new-town/peterborough/. Accessed 09 Jan 2023.

Vivid Economics. 2017. *Natural capital accounts for public green space in London,* Report prepared for Greater London Authority, National Trust and Heritage Lottery Fund, October 2017. London: Vivid Economics. https://www.london.gov.UK/sites/default/files/11015viv_natural_capital_account_for_london_v7_full_vis.pdf. Accessed 29 Jan 2023.

CHAPTER 5

Municipal Care: Public Parks and the Governance of Place

Abstract This chapter examines the crisis for England's public parks which are predominantly owned and managed by local government, and which have been subject to budget cuts and changing governance arrangements from the late twentieth century to present. It identifies the underlying causes of these changes to the political economy of park-making and management, as ideological, an argument supported by the continuing lack of statutory duty to maintain access to these public spaces, despite evidence of their significant contribution to public good through place-making, economic development, health and well-being. The chapter considers the recent calls for reform and central coordination of land use and access in England and Scotland and returns to the six ecosystem case studies of the Understanding Everyday Participation (UEP) research to explore perception and experiences of place governance through everyday participation in parks and green spaces in these locations.

Keywords Municipal government • Public parks • Statutory duty of care • Place governance • Business models • Re-thinking parks

© The Author(s), under exclusive license to Springer Nature
Switzerland AG 2023
A. Gilmore, *Culture, Participation and Policy in the Municipal
Public Park*, Palgrave Studies in Cultural Participation ,
https://doi.org/10.1007/978-3-031-44277-3_5

135

INTRODUCTION

The development of municipal public parks in England across the nineteenth and twentieth centuries, discussed in Chap. 2, involved the making of new national and local policies to rationalise the use of public funds for their establishment and upkeep. The arguments put forward for public parks for the health, well-being and productivity of the nation, to Select Committees and through legislation, successfully put into place a mandate to dedicate specific property in public trust, as a common resource that would be stewarded by its owners and users in perpetuity. This policy formation, driven by the relations of expanding capitalism, industrialisation, shifting class structures and leisure practices sought out new spaces for cultural governance and the legitimation of authority over crowded, unhealthy cities and their populations. Along with free libraries, museums and other civic institutions and public realm improvements, public parks were part of a new municipalism and their management and coordination influenced the development of its governmental structures and regulatory processes. Whether it was the gifting or purchase of emparked land of manor houses which had been rendered semi-public by their enclosure by urban housing and manufacture or the repurposing of green space from previous use such as cemeteries or wasteland, the making of urban parks which held in place the rights of access for all publics came with a legacy of significant obligations for the local state.

To stay public and to remain useful, parks need protection, conservation, ongoing management, and care, requiring resources and decision-making from their owners and stewards. Within the UK responsibility for an estimated 27,000 public parks and greenspace remains with local authorities. However, the post-crash austerity agenda of the UK government has led to a decrease of 30% local authority spending power, following ongoing cuts in grants from central to local government amounting to an estimated 56.3% real-terms reduction (Rex and Campbell 2022, p. 26). Part of a portfolio of non-statutory public expenditure within culture and related services, public expenditure by local authorities on parks has fallen by 8.5% over the last 11 years, and whilst their potential to raise income has had a marginal overall increase of 15%, there is still clearly the need for significant grant income (see Fig. 5.1).

Real-term cuts on parks and greenspace are estimated at £330 m less per year, amounting to nearly 25% decrease over a decade (Martinsson et al. 2022) and significantly impacting levels of service provision. Whilst

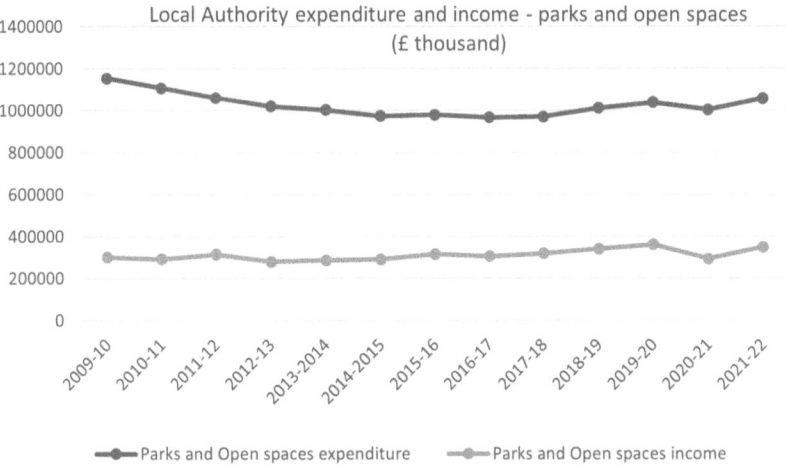

Fig. 5.1 Local authority expenditure and income—parks and open spaces, England only. (Source: DLUHC/MHCLG 2022)

there is a jump in overall expenditure on culture and related services in 2020–2021 (see Fig. 5.2), this was largely due to the subsidy of local authority-managed amenities such as sports and performing arts centres whose business models fell apart during the pandemic (Walmsley et al. 2022).

The impact on statutory and non-statutory services delivered by local authorities is viewed as a part of the general retraction of the neo-liberal state by the political left, described as a case of 'staggering austerity' that has a disproportionate impact on society's more vulnerable and marginalised people before moving on to areas of service that are both statutory and deemed essential that the wider public care about (Ryan 2017). Whilst the effects of local authority expenditure cuts on arts and culture are vastly differentiated across the country (Rex and Campbell 2022), the decline in public parks has been universally decried, with successive public petitions and media outcry, and a cross-party Select Committee on their future, discussed below. Meanwhile, there remains no national agency that has responsibility for municipal public parks, nor in England and Wales is there any statutory duty for their protection. The focus and objective of this chapter is to consider the politics and policies of municipal parks in the UK within this context, and ask the following questions:

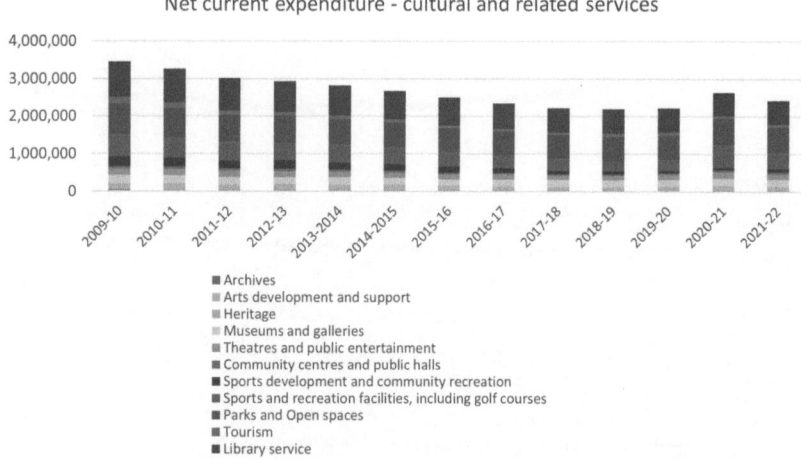

Fig. 5.2   Local authority expenditure: culture and related services, 2009–2022 England. (Source: DLUHC/MHCLG 2022, General Fund Revenue Account Outturn RO5 Cultural, Environmental, Regulatory and Planning Services)

- What are the policy rationales for preserving public parks in the twenty-first century?
- Who is involved in their management and stewardship?
- How does the management of parks underpin and sometimes undermine the possibilities for cultural democracy?

To do so, I explore the recent debates and discourses within policy literature which articulate the values and rationales for maintaining public parks at both national and local levels. I also look at how these manifest empirically by returning to the *Understanding Everyday Participation* (UEP) ecosystems case studies, introduced in Chap. 4.[1]

The chapter has the following structure. The first section considers the political economy of urban public parks. It traces current issues of resource management and sustainability back to the establishment of parks within public trust and the transfer of land ownership to local government. This, I argue, creates a paradox of value for parks in which the public value of municipal parks expressed through evidence of participation and extrinsic benefits contradicts with the unrealisable market value of urban land. As there is no statutory duty for local authorities to look after their municipal parks, the gathering of evidence is encouraged to make the political case for public investment within finite (and decreasing) state funds, and to defend them from enclosure and depletion. The paradox is that the wealth of evidence for the public values that parks generate is not rewarded by their safeguarding, yet the 'business case' for parks is weak, their asset value is low, and commercialisation would lead to their enclosure, privatisation and consequent decline in public value. I go on to outline how these values are instantiated within local and national government policies, which frame their potential contributions to health and well-being, and social infrastructure.

The second section considers the discourses on public space management and the insights they provide into affective relationships between what Sennett (2018) calls the *cité* and the *ville* found within residents in Stornoway and Peterculter in Scotland, Peterborough, Dartmoor and Manchester in England, which were case studies in the Understanding Everyday Participation project. These case studies were introduced in the previous chapter, where I look at the participation narratives that articulate the value and meanings of access to and use of parks and green spaces and the social and cultural practices that take place in them. Here, I look at the perceptions of responsibility, ownership and protection of parks and open greenspace that emerged in interviews, ethnography, workshops and practice research. These reveal the inherent relationships between land, community and participation embedded in national legal frameworks, which are manifest within local governance of place and public space and dictate who has the right to land usership and extraction. I argue that these discourses which surround these national-local policy flows represent not only the policy challenge that municipal public parks pose, but also the distinctive affective relations between *cité* and *ville* within this range of place-types and cultural ecosystems, which might also be considered their 'local structures of feeling' (Taylor et al. 1996). Finally, I return to the politics of parks management in Manchester and Salford, the birthplace of the municipal public park, to consider the continued threats to public space within the neoliberal city.

BUSINESS MODELS FOR PARKS AND THE PARK
VALUE PARADOX

In North Manchester on 22 August 1846, at Queens Park, one of the three first municipal public parks in England opening that day, Mark Philips addressed the assembled audience of tens of thousands. Philips, the first MP for Manchester who was about to retire after 15 years of service and who gave his name to one of the three parks, urged the crowd to 'keep this park properly as though it were his [sic] own' (Ruff 2016, p. 28), to encourage friends and families to identify areas that they particularly valued and look after them, and to alert others who they observed to be careless to also get involved in stewarding this new green resource. The mayor of Manchester put this more bluntly: 'Mr Philips puts me in mind that I should have told you this park is now handed over to you, and that we have done with it' (Ruff 2016 p. 29). This instruction to the people to police and look after their own parks neatly illustrates the fault-line on which the business models for public parks have been precariously balanced since their inception. Although not all parks began under their stewardship, local authorities became the default landowners of parks that came into their hands, passing from private hands into public ownership and bringing with them the vestiges of failed financial models. Amongst these were hybrid models comprising public parks and private gardens which retained access for the previous landowners whose sold-off property subsidised their creation (Layton-Jones 2016).

The gifting and rescuing of parkland from philanthropists, trusts and private developers joined the new parks developed under public and political pressure. As Conway's (1991) history of municipal parks reveals, even when instigated by public solicitation and with subscription funds raised for park landscaping and design, the opportunity costs for presenting a beautiful public realm to the people was rarely offset by protecting the projected costs of maintenance. The handing over of finished parks for the benefit of the public in perpetuity contributed to a legacy of unsustainable responsibility accompanied by the enduring expectation that the authorities should look after parks on behalf of the people. Placards displayed during the launch of those first three municipal parks stated: 'This park was bought by the people for the enjoyment of the people and committed to the people for protection'; as Ruff comments, these were words that would come back to haunt the public parks committee established in

Manchester, Salford and elsewhere across the decades to the present day (2016, p. 28).

As described in Chap. 2, local authority development of new public parks continued apace in the nineteenth century, accompanied by declining private development, facilitated by Open Spaces Act of 1887, which increased the rights of local authorities to raise the funds necessary to acquire and maintain land (Layton-Jones 2016). This led to fewer private projects amid the growing expectation that local authorities would fund and manage parks. In such an environment, developers withdrew from park provision entirely, leaving local authorities with the combined burden of managing inherited sites and developing new parks simultaneously. By the end of the First World War, local authorities across the country found themselves obliged to receive under-funded parks in various states of disrepair. In her substantial history of municipal parks, Conway identified 23 public parks which, by 1991, had passed into local authority control from private companies, trusts and the royal estate (Conway 1991).

Over the second half of the twentieth century as culture and leisure patterns changed, municipal parks became the arena for an increasing range of (built) recreation amenities with corresponding increases in local government responsibility and expenditure. The look, feel and popular success of urban parks was affected by other external events: Harding highlights the deleterious stripping of park infrastructure for War effort, as the removal of iron railings and gates opened up park spaces and laid them bare to anti-social behaviour and damage. Ultimately, a combination of factors affecting local government operations in the 1970s led to diminishment of their long position as the major source of civic pride. These included local government reorganisation in 1974 which demoted park management roles in favour of a broader portfolio of cultural and leisure offer and estate management of sports centres. They also included the major shifts in policy and rifts in the relationship between central and local government following the advent of the Conservative Party rule in 1979, which brought in private outsourcing of public services through Compulsory Competitive Tendering whilst slashing central funds to local authorities, forcing parks funding to give way to statutory requirements (Harding 1999).

The decisions to protect and maintain parks in England and Wales are bound by political mandates rather than statutory frameworks, based on valuation of their returns on investment and their popular appeal within the lexicon of New Public Management, which requires the creation of 'public value in such a way that what the public most cares about is addressed effectively and what is good for the public is put in place' (Bryson et al. 2014, p. 447). At the heart of the issue for many local authorities is the value that parks present on their finance ledgers: as many parks under their management were either gifted or were residual common land, to be held in public trust in perpetuity, they are marked on balance sheets as having a £1 asset value and seen as low priority for investment compared with other local government-owned assets (CabeSPACE 2009). Continuing decline in the previously high standards of presentation and care in the decades that followed revealed the public park's position as a 'soft target' within local authority budget cuts, precipitating campaigns which aimed to make a better case for public investment (Crowe 2018). The debates near the turn of the millennium, as a New Labour government came into power, foreshadow contemporary concerns for public parks, as they remain non-statutory and vulnerable to the austerity financing of local authorities by a Conservative central government. As before, these concerns triggered a re-evaluation of public parks to inform their advocacy, discussed in more below; in the 1990s, these campaigns were fought by voluntary and professional groups who lobbied for public funding, a case which was primarily made—and won—with reference to their historic and heritage values.

This pivotal moment came in 1995 when the newly formed Heritage Lottery Fund (HLF)[2] turned to the plight of urban municipal parks and announced its intended and ongoing commitment to the 'one area of our popular heritage where we can really make a difference to people's lives' (Durham 1995, p. 3). Tandy (2019) identifies the root explanation for this 'new heritage practice' (p. 203) as a combination of the publication of two consultancy reports with the need to address public relations missteps associated with the controversial acquisition of the Winston Churchill papers by the arms-length national lottery distributor earlier in the year, which had caused discomforting accusations of elitism in the HLF's early investment choices. This effectively reframed the cause of parks decline from one of poor local management to national heritage crisis. The two reports contributed to this shift in different ways. The first, *Public Prospects: Historic Parks under Threat* (Conway and Lambert 1993), commissioned by the Victorian Society and the Garden History Society and prompted by the Review of Royal Parks, identified the heritage value of parks through their place in landscape design history when transferred from aristocratic private estate to urban contexts, and urged the protection and reinstatement of original planting designs and historic features. The second report, *Park Life: Parks and Social Renewal* (Greenhalgh and Worpole 1995) funded by left-wing thinktanks, Comedia and Demos, who were pitching policy schema in anticipation of the forthcoming political change in government, focused on parks as a source of urban renewal, community cohesion and 'sense of place'. Thus, a new Urban Parks Programme was launched by the HLF in early 1996 in Weston Park, Sheffield, which combined populist appeal of parks with heritage valuation as a rationale for lottery funding to present a new source of regeneration funding for blighted local governments.

As Tandy (2019) documents, this was no straightforward exercise not least for the heritage experts of the HLF who lacked experience in managing public parks or assessing which of the many aspects of their restoration should be funded. It was a considerable task to work through the ongoing deterioration in so many locales, to select projects that met eligibility criteria, which included community participation and engagement with local government, and to rationalise this expansive new heritage practice. Whilst research evidence could help demonstrate the significance of design and support the conservation of historic features such as planting, monuments and built infrastructure, to fully rejuvenate parks for popular appeal, basic visitor facilities such as toilets and cafes needed updating (Bramhill and Tibbatts 2002). As a result, the HLF:

> committed far more money than intended to tackle the massive backlog of repairs to essential park infrastructure. It also tried to address the causes of decline in partnership with local authorities and other bodies in the field, including loss of management structure and skills, lack of political support and understanding, and the dearth of relevant data concerning parks. (Crowe 2018, p. 61)

The Urban Parks Programme was followed by a follow-up programme, Parks for People, which was jointly funded by the Big Lottery Fund and ran throughout New Labour's period in government, from 2006 to 2012. The sequential funding of park restoration and rejuvenation established the HLF as major convenor of the management of parks nationally through the resulting partnerships with local government, sector and community groups and through the collection of data and evidence of value of investment, such as the two State of the UK Parks reports in the mid-2010s (HLF 2014, 2016). These presented evidence of the scale of the problem, highlighting both public perceptions of parks' decline and the reality of service and workforce cuts within local authority parks departments; they also demonstrated the popularity of parks as sites of everyday participation, finding over 37 million people, more than 57% of the UK population, regularly use parks in the UK (CLG 2017, p. 6). The findings supported further partnership initiatives aimed to produce new business cases and models through R&D in pilot project areas, such as Rethinking Parks with Nesta (2018), and the Future Parks Accelerator launched in 2019, with the National Trust and the Department of Levelling Up, Housing and Communities (NLHF n.d.). However, despite £950 million of National Lottery funds distributed over 25 years (NLHF n.d.), the need for continuation funding and the quest for sustainable business models continue. Unlike other heritage artefacts, as organic eco-systems public parks require continual management and maintenance, and arguably the support of the HLF increased the need for revenue funding (Tandy 2019), creating further expenditure demands to be picked up by local authorities when capital and project funding programmes expire.

RE-INFLATING THE LUNGS OF THE CITY: NATURAL CAPITAL AND EVIDENCE OF VALUE

These continuing concerns were the topic at the heart of the Communities and Local Government Committee on Public Parks that took place over 2016 (CLG 2017). The context for this inquiry was the continuing austerity brought in by the coalition government in 2010 in an attempt to address public borrowing, which led to ongoing cuts by Treasury to local government budgets, enforcing increased competition over decreasing resources for locally delivered public services and after six years bringing even those protected by statutory duty to near-breaking point (Ryan 2017). The cross-party CLG Committee considered three questions: why parks matter, what are the main challenges for their management, and what policy and funding models might secure their future sustainability? Answers to the first question were found in the evidence of the social, environmental, community, health and well-being benefits of parks submitted to the inquiry, leading the Committee to name parks 'treasured public assets' (2017, p. 45). The evidence submissions rehearse many of the arguments presented by witnesses and in open letters to the 1833 Select Committee on Public Walks that spearheaded the original campaign for the municipal public parks (as discussed in Chap. 2), albeit with more sophisticated data capture and economic modelling. The Committee had commissioned its own survey of 13,000 respondents on their perceptions and participation habits, which found, for example, that nearly 90% of respondents feel their local park has a very positive effect on their health and well-being, over 45% visited two or three times a week, and nearly 25% every day (CLG 2017). Added to these survey data, which included open text and descriptive statistics, were many qualitative and quantitative accounts of value within the 400 written evidence submissions and oral evidence sessions. These included survey data on increased frequency of physical exercise linked to correlated improvements in social, cognitive, cardiovascular and immune system functioning, narratives of community integration and social inclusion, and evidence of mitigating effects on flood risk, pollution and urban overheating and other 'ecosystem' services such as biodiversity and carbon storage.

The CLG Committee also critically reviewed further research instruments for their potential to provide robust evidence that might meet the demands of the government's Treasury Green Book and prove useful tools in local authority advocacy and case making for further funds. These methodologies aimed to quantify values to public health and ecosystem services to show cost-benefits parks contribute to broader agendas in economic terms. These, they argued, might also be used to lever private monies, for example, from water companies in return for environmental protections, or to calculate cumulative social returns on investment, recognising the 'traditional accountancy methods' of operational costs and land values were not adequate (CLG 2017, p. 21). Whilst the CLG inquiry warned of a lack of capacity on the part of local park managers to undertake the complex valuations such evidence-making requires, they still welcomed 'new models which are emerging to help assess the value of parks' broader contributions in a more nuanced way' (2017, p. 22). More evidence could help target investment and support cost-cutting, and ease in social investment and volunteer value-in-kind through community contribution to the upkeep of parks.

Crowe (2018) similarly identifies the 'pricing' of the multiple benefits of green spaces through 'natural capital' and 'social return on investment' methods, promoted by government as a step-change in evidence-based decision-making and creating headline assessments such as healthcare cost savings of £950 million for Londoners, and calculations of £34 billion across the UK of health and well-being benefits (p. 63). However, significantly, she also notes the lack of traction this compelling evidence seems to create when it comes to debates in central government which appear to disconnect the method from any meaningful contribution. This is since there is no market mechanism to actively extract the quantified values of natural capital as some sort of return, and:

> even if this mechanism was apparent, local authorities are not in a position to divert resources from other priority areas, even if they accept some sort of 'value for money' or 'opportunity cost' rationale. The resources are simply not available. (Crowe 2018, p. 64)

The strength of advocacy for parks via research evidence of their value can be seen by the huge volume of reports published at and since the time of the Committee: over 60 reports detailing economic value, environmental, community, health and well-being benefits are promoted by the Parks Alliance, the sector coalition that formed in 2014 in response to the funding crisis (The Parks Alliance n.d.) and there were nearly 400 submissions of evidence to the inquiry itself. The clamour for such evidence is less a strategy for success than a symptom of the policy thinking emerging from what Crowe names a 'creeping neo-liberalism' (p. 63) within post-New Public Management austerity (Wells 2018). This requires technocratic measures of accountancy and evidence-gathering to support policy rationales on the one hand, whilst on the other, the institution of central government cuts to local authorities leave no option to disinvestment other than market-based solutions, mirrored by the language in policy literature of 'creative management' and 'park innovators' to take the burden off the state (Nesta 2018). This is also the case for parks trust models where charitable organisations are established to take over financing and management by raising endowments and developing complex financial and legal arrangements through licensing and lease of land to promote income diversification, a strategy that has been in place in the new town of Milton Keynes since 1992 (The Parks Trust n.d.), and recently been adopted in Newcastle (Urban Green Newcastle n.d.). The CLG Committee heard from CEO of the Milton Keynes trust of the 'innovation' from having a 'single purpose', 'free' from the responsibilities of local government (David Foster, cited in CLG 2017, p. 52) and whilst some inquiry participants voiced concerns that trusts erase democratic accountability outside of local authority control, these were countered by the option proposed by the head of the Land Trust to experiment in 'soft ownership', encouraging community involvement on the basis that:

> [the park] feels like it is mine; legally and technically it might not be, but it feels like it may be mine. (Alan Carter, cited in CLG 2017, p. 53)

The recommendation of the Committee to government was for the continued pursuit of alternative, non-traditional management models for parks, with the proviso that there should be clear guidance on accountability structures and further change management funding for the embedding of new practices learnt from pilot projects such as Rethinking Parks. The CLG Committee members noted the absence of coordination of policy for public parks at a national level; despite the existence of arms-length and government-sponsored bodies for natural environment (and indeed arts, culture and heritage) there was an identified need for a dedicated agency that can 'get parks and green spaces out of this very marginal local authority leisure place, and right up to being really important national strategic infrastructure' (Julia Thrift, Town and Country Planning Association, cited in CLG 2017, pp. 62–63). Whilst sector bodies like the Parks Alliance provide a 'voice for parks' nationally, they can only lobby and advocate seemingly often to deaf ears.

A STATUTORY DUTY OF CARE

The CLG Committee stated that they had listened to the concerns of the 322,000 petitioners who had signed a call for the statutory protection of public parks issued by 38 Degrees (CLG 2017, p. 7) and noted that these repeated concerns of the ODPM: Housing, Planning, Local Government and the Regions Committee (ODPM 2003) which had recommended statutory duty, albeit during mid-New Labour term in office and in the context of a central government funding for environmental services. However, following the inquiry of 2016 this recommendation was not forthcoming. Instead, the Committee warned of the challenges for monitoring and auditing local authorities who would need to confirm they had delivered on the duty. Rather than taking on this challenge, local authorities should work collaboratively with Health and Well-being Boards, through joint parks and green space strategies, to articulate the contribution of parks into wider local objectives. Local partnerships should consider parks as a portfolio, and work strategically with all parks and their communities of interests, planning in alternative business models for sustainability. This, it might seem, could only increase the complexities of monitoring and generating the evidence for different policy rationales at local level.

Considering the statutory duty question, Dickinson et al. (2019) examine the impact of austerity measures on local government in exacerbating a downward spiral for parks in the UK, and extend this analysis to a global crisis in greenspace funding, citing examples from China and America. They similarly identify the multiple narratives on the societal values parks produce and speculate that greenspace 'potentially fills the gap in the market' created by other types of everyday 'Third Place' such as public house, library and sports facility (Dickinson et al. 2019, p. 123) which are under threat from funding cuts and entrenched deprivation. The use of the term of the market is unfortunate, or perhaps deliberate, since it highlights the problem at the heart of the park paradox: it is the lack of opportunities for extractive value that parks present which prevents their economic sustainability, whilst their market exposure and quest for income generation commodifies and encloses them as pseudo-public spaces. Indeed, the authors comment on how even the encouragement of volunteer involvement in park management might threaten their public status—suggesting that without local authority control there could be conflicts over ownership, as different stakeholder groups might illegally cultivate greenspaces through 'guerrilla gardening' (Dickinson et al. 2019, p. 124).

To see how this conundrum might be remedied, Dickinson et al. use a doctrinal research methodology which plumbs available case law, statutes and legal rules, to model the transference or 'recasting' of existing legal authorities in other contexts. The current basis for regulation in the UK is founded on accessibility, quality and service standards developed within the policy and grant assessment criteria for national non-departmental public body Natural England, which forms the basis for the Green Flag Award scheme. However, this scheme has no legal basis for implementation, nor does it compel local authorities to provide any legal redress for any breach of these standards. It is complex, as the CLG Committee warned. There are bye-laws that can be invoked for anti-social behaviour which were empowered by the Local Government Act 1972, Section 235 and can be developed and interpreted locally and regulated through fines and convictions; older Acts such as the Open Spaces Act of 1906 and the Public Health Act of 1875 provide local government with powers to purchase and promote greenspace for leisure and recreation, and planning frameworks provide some guidelines for conservation and protection from development. However, these are relatively blunt instruments that rely on discretionary rather than the statutory obligations that exist for over 1000 other areas of local authority governance, such as housing, council tax and so on, as was called for by CLG Committee participants, and within other previous consultation (Lee et al. 2015). These requests to central government to make parks statutory service provision have failed, however. Dickinson et al. (2019) argue that the challenge is to demonstrate potential success in ways that are less abstract: this is where their doctrine research methodology comes into its fore, as the article recasts the Scottish Land Reform [Scotland] Act 2003 onto the case of England and Wales.

This Act contains specific guidance that provides a turnkey solution to the tensions between national imposition and local ownership and control, in that it requires the statutory duty to uphold access rights which enable

> all members of the public to enjoy the countryside and to take part in informal recreation [...] Local authorities have a key role in ensuring that these access rights are facilitated on the ground. (Scottish Government 2005, cited in Dickinson et al. 2019, p. 127)

This applies to land even where local authorities do not have responsibility and includes the obligation to ensure access is compatible under equality laws, whether for access for those with disabilities or other aspects that might withhold equal access. The duty also has the provision that local authorities are not required to do things that would jeopardize their ability to fulfil other duties, so when faced with planning applications for developments on land with existing access rights, they can both grant consent whilst attaching conditions for maintaining reasonable access. Furthermore, the authors argue, this duty engenders local engagement with community forums to ensure all stakeholders are represented, and the Act also provides means for local authorities to negotiate liability and consider conditions locally, to temporarily allow activities that might prevent access rights permanently if not regulated. For Dickinson et al., the recasting of this duty for England and Wales would be a far less uncertain option than the push for alternative models suggested within the CLG recommendations, providing 'both a carrot and a stick to local authorities to achieve their green infrastructure' (2019, p. 132).

Everyday Management of Parks and Greenspaces

The questions of how municipal parks are funded, regulated and protected continue to be of concern at a national level, despite repeated case-making, lobbying, parliamentary debate and select committee inquiries. They manifest locally in the frustrations of local authorities and their search for alternative income streams, management strategies and mechanisms to engage local communities in park advocacy and stewardship. These tensions and debates also emerged within the Understanding Everyday Participation (UEP) research, which took place over a five-year period leading up to the CLG Committee. As described in Chap. 4, the research fieldwork involved household interviews, ethnography, and data mapping in consultation with local authorities, cultural institutions, and community groups, in six case study ecosystems. Each case study presented different socio-economic histories and political trajectories, governance arrangements, industrial and spatial topographies, demographic patterns, and cultural identities. These make up what Sennett (2018) understands as the *ville*, or planned environment, which is caught up in a contingent but not necessarily symbiotic relationship with the *cité*, or lived experience, sense of and attachment to place. Good urbanism finds ways to bring these two understandings of place together, Sennett argues, and he proposes architectural and planning 'open' forms which support participation, synchronicity and social mixing.

In this book, I argue public parks are spaces where, through everyday participation practices, the affective relationships between *ville* and *cité* are played out and brought together. Parks possess many of the design qualities that Sennett proposes are 'open form': different zones separated by landscaping and built infrastructure encourage synchronous rather than sequential activities. Low railings and open vistas mean that parks are porous, promoting visibility and social contact without imposing it, mediating conviviality (Barker et al. 2019) and side-by-side living. Of course, the planned environments of the *ville* are not shaped solely from hard infrastructure but also comprise the soft or 'social infrastructure', the communal spaces, public services and processes that support social practice, meaningful relationships, trust and reciprocity (Kelsey and Kenny 2021). They also, I would argue, require spaces for dissent and resistance, and parks are such places. As discussed in earlier chapters, people take part in emplaced practices of locational citizenship (Di Masso 2015) through the ways they resist, contribute, shape and enact the public-ness (and policy-ness) of parks and greenspaces, even when it is simply walking or cycling through them (on or off the path), dwelling longer for a picnic, to volunteer in community gardening, to watch people or play sports. By looking at these routine, everyday social practices and the values attached to them, we can investigate the practicing of citizenship and explore the *cité* or lived experience of a place.

The *ville* of the UEP case study locations was revealed as interviewees talked about their perceptions of local governance, issues and concerns about the stewardship of public parks and greenspace, the factors that affect or create opportunities for participation in their local neighbourhoods. For example, the experimental planned environment of Peterborough originates in the 1960s, influenced by the Garden City movement as new townships were added concentrically around an historic centre, ringed by relative affluent peri-rural settlements. Small parks and recreation grounds pepper these residential townships with further greenspace providing pathways and routes around the conurbation and interconnecting the larger parklands to the southwest of Mitton Park, Ferry Meadows and Thorpe Park which frame the waterways and lakes along the course of the River Nene. The topography of the city is reflected in the navigation of public space within everyday participation, for example, for this interviewee:

It's almost as though [Peterborough] grows areas and then they grow things that happen ... once they're there and established you go to them ... but it's almost as though places come without that designed and then somehow they ... form themselves ... .I think every time you do, you know, you move in one direction you sort of change the focus of everywhere else. (Peterborough participant)

They went on to say:

This is the community difficulty thing in a city like Peterborough, where there are a lot of things that are across the city rather than things happening in a small area ... must be things going on but to be honest I don't know now what they do have going on. (Peterborough participant)

The promotion of parks and greenspace as places for participation has been under review by the Cambridgeshire and Peterborough Future Parks Accelerator, one of the pilot projects developed following the CLG

Committee and funded by the National Lottery Heritage Fund, the National Trust and the Ministry of Housing, Communities and Local Government. Partnering with seven local authorities and the sports development charity, Living Sport, the project promotes parks, their events and resources, providing details of their locations and amenities within the region's 200 parks and greenspaces via an interactive map, and a segmentation model, to aid navigation which broadly classifies parks by size, types of amenities and natural and artificial features (see Fig. 5.3).

The segmentation also informs open data mapping developed to give local authorities across the study area baseline standards for assessing their park provision (Cambridgeshire Council Council n.d.). At the same time, the local authority for Peterborough identified the need to further classify

Park Type	Definition	Example
Country Park	A large park setting providing a wide range of recreational activities including outdoor sports facilities, play equipment and informal recreational pursuits, nature trails, cycle routes, formal picnic areas, facilities refreshment and toilets.	Nene Park Trust
Neighbourhood Public Parks & Gardens	Park setting providing a variety of recreation activities, including outdoor sports facilities and playing fields, play equipment, informal seating areas and walking (e.g recreation grounds).	Cherry Hinton Hall
Natural & Semi Natural Open Space	Park setting including a range of natural features; Woodland, scrub, grassland, heath or moor, wetlands open & running water wastelands, wetlands (e.g. nature reserves).	Somersham Local Nature Reserve

Fig. 5.3 Taxonomy of parks in Cambridge and Peterborough. (Cambridge Open Space n.d.)

new parks and open space with more granular detail on the distinctions between county, community, and neighbourhood parks, for planning purposes, creating a new second tier of urban park community parks, including Urban Squares, Pocket Parks, Sliver Parks, Courtyards and Connecting Links (The City of Peterborough 2019a). These recognised the growing urban density of Peterborough: although the city has an above-average provision of open spaces for public use, culture and recreation, much of which has natural heritage assets, such as wetland, lakes and woodland, there is a below-standard quantity of suitable parkland within areas of increasing residential density. The resulting assessment framework provides quantitative standards including recommended hectares per population, quality scores, minimum requirements for design and construction, and a formula for calculating 'park equity', combining criteria for accessibility, quality and inclusivity (the extent to which all residents can access parks and open spaces) plus density and income levels of proximate populations. These provide the local authority with the necessary technocracy for case-making and targeted investment, for example, in the rejuvenation of existing parks, acquisition of parkland and the designation of certain parks as 'preserves' and 'reserves' to protect their natural heritage from development as the city continues to grow (The City of Peterborough 2019b).

Whilst 60% of the land area in Gateshead in the northeast of England is rural in character (Gateshead Council 2013), the UEP case study research focused on an urban corridor which stretches away from the quayside where the city meets the Tyne, overlooking its sibling city Newcastle. This is the location of Gateshead main shopping centre, but also of the Baltic Contemporary Arts Centre and the Gateshead Sage music centre established through Lottery funding at the turn of the millennium, the outcome of long campaign for capital investment in arts and culture institutions by the local authority, the regional arts board and other local advocates (Gibson 2019). The corridor includes distinctive areas characterised by demography, street patterns and housing stock: at its north, the wards of Saltwell and Bensham have poorer households living in terrace, flats and ex-Council housing with a declining set of heritage buildings and features, and to the south lie the more affluent, middle-class areas of Low Fell and Chowdene, with older populations in detached houses and Victorian villas, further from the industrial areas around the Tyne (Gibson et al. 2014). In the middle of the corridor is Saltwell Park, the very large Victorian park originally named the People's Park and established in 1876, as the owner of the land, stained glassmaker William Wailes sold his estate to Gateshead Corporation, when he ran into debt (Historic England 1985). Extended by the local authority in 1920 and including 11 listed buildings, a lake and a maze, the park deteriorated significantly over the twentieth century and was the recipient of £10 million HLF between 2000 and 2005 funding to restore it to its original landscaping. Since then, the park has won consecutive Green Flag awards and was named Britain's Best Park in 2005, and Civic Trust Park of the Year the following year (Gateshead City Council n.d.).

The UEP interviewees also recognised the importance of investment and management of parks in their neighbourhoods. They spoke of the significance of parks within their broader environment, and health benefits they offer as spaces for exercise. For example, one interviewee, who had suffered from chronic fatigue syndrome earlier in life, talked not just about individual health benefits but also the importance of parks to their broader environments:

> I think if you have your surroundings improved everybody feels a lift. I think the fact we've got Saltwell Park is such a plus thing for Bensham, at least you've got somewhere to go and walk that's pleasant and nice and nice views and that. It annoys me when they talk about getting healthy, well, go for a walk. But where do you go for a walk in some areas? (Gateshead participant)

This interviewee contrasted the park with the dowdiness of the town centre which had until recent 'improvement' been a social space:

> It seems that the town centre used to collect a lot of people who you might consider almost misfit. But the market was—, although it was a dump, the market at least had a place where they could—, a lot of them just used to—, 'cause I worked there at [mumbles], I had like a—, hiring out mobility scooters. But for a lot of them it was their social. They'd be there in the morning then they'd go away and they'd be back in the afternoon, and they'd just be walking around and talking to each other. I don't know where they are now because clearly, it's not the same in the town centre because it's just like a normal town centre. (Gateshead participant).

These comments seem to recognise the innate contradictions within the 'improvement' of places that offer suitable spaces for gathering, synchronous activities and serendipitous encounter, the open form properties of places such as markets and parks, alongside the need to promote participation opportunities on an equitable basis.

In rural Dartmoor, in the county of Devon, southwestern England, the fieldwork took place across peri-rural settlements Moretonhampstead, South Brent, Princetown, Postbridge, Heathfield, Widecombe which formed a ring on the edges of the moor distanced from larger conurbations, such as Exeter, Plymouth and Totnes (see also Fig. 4.2 in Chap. 4). The research involved household interviews and ethnography in Moretonhampstead and Buckfastleigh, which are characterised by relative socio-economic affluence but also hybridity, as rural places which have aspects of urban living (commuting to work, for instance) whilst effectively surrounded by countryside (Schaefer et al. 2017). Their planned environment is dominated by the Dartmoor National Park which is also flanked by five Areas of Outstanding Natural Beauty in Devon, geographies which have statutory protection overseen by the relevant local authority, and in the case of the National Park, planning authority, designated by Natural England.[3]

The research identified attitudes to the use and protection of the surrounding landscape, which gave insights into the layering and hierarchies of local governance, which includes the National Park authority, parish and district councils and Devon County Council, as well as community groups and third sector bodies. Parks and greenspace reachable by walking in the towns were observed as important but under-resourced, poorly managed, and needing community or third sector intervention due to a lack of investment from the local authorities:

> They were taking money away from everything recently … Green space, parks, well it seems like they're always fundraising from other places to do the parks and things. (Dartmoor participant)

This interviewee referred to funds raised by Buckfast Abbey to support its local gardens, and also the number of open air swimming pools, skate parks and playgrounds, such as Victoria Park in Buckfastleigh, which are managed by charitable trusts and operated by volunteers, reflecting a strong local culture of community participation.

By contrast, the competence of those with decision-making powers, both local and national, was often questioned:

I think they could probably do with directing a bit more attention to the people who live here, rather than just trying to finance business. You know, I think the welfare state side of things could probably do with a bit of attention, but we are under a Tory government, so you know ... in rural areas I think it's quite easy to feel like we're being forgotten about and that we don't matter. (Dartmoor participant)

These communities appeared to find themselves situated on the edge of state policy that focused on protecting landscape rather than its surrounding peri-rural communities. There was a sense that local interests were neglected in favour of others, in spite, or perhaps because of the proximity of the dominant National Park.

But I think they're there at Park, which is what they're called, they're there at Park [Dartmoor National Park authority] in their little ivory towers and I think they have very little knowledge of how to develop but at the same time preserve. (Dartmoor participant)

This same interviewee stated:

A lot of Dartmoor, especially the peripheral areas, 'cause we're almost in the national park here, are neglected and that care and protection instead was offered to tourism hotspots like where the royalty go for the summer or if they're near a very big conurbation. (Dartmoor participant)

This ambivalence was reflected in the findings of Schaefer, Edwards and Milling (2017) who led the case study research in Dartmoor, and who observe the targeting of 'investment and service provision in urban hubs, leaving contrasting rural areas doubly disadvantaged' (p. 47). Their study identified high levels of cultural participation amongst local people but not in state-funded activities, not least as state funding simply did not reach these communities. They noted that this participation helped their subjects to negotiate the 'increasingly ambiguous senses of place' (p. 54) and the hybridity of their locations, placing them in and outside of the urban and the rural landscape.

The evaluation of local authorities, their care for the environment and contribution to social infrastructure was also manifest in the two Scottish case study ecosystems. In Peterculter, Aberdeenshire, a peri-rural coastal village with relative affluence from its oil industry residents, but working-class roots and pockets of social housing bringing heterogeneity (Miles and Ebrey 2017), there were signs of the retraction of local government and creeping privatisation of services:

> They do a good job with the transport but that's not the council, that's a private group isn't it? The transport's very good, the parks—, I can see a slowdown in the—, like leaf cleaning, they don't bother so much, there's a kind of reduction in these services. Again, they've done a disastrous job with Union Terrace Gardens, they hadn't got a clue. They've done a disastrous job with the Spring Road, they've done a disastrous job with our road system, it's—, so they've done some good things but they've—, they're cutting back on funding for like arts and environment, so. (Peterculter participant)

The interviewee continued:

> The local people should step in then to fill that gap. But if it's a service that the people are paying through, it's part of the service and they're paying through their council tax then they should expect and demand that service, yeah. (Peterculter participant)

The expectations for communities to step in, whether to fill the gap or demand that local authorities' duties are carried out, are enlightening. With few formal institutions or cultural funding, the social infrastructure of Peterculter is provided by its community, and Miles and Ebrey (2017) argue the 'moral currency' (p. 63) of the volunteering imperative and accompanying narratives of mutuality and reciprocity form a central plank of the 'social imaginary' (p. 62) of everyday life in the village.

It may seem strange to use the terms *ville* and *cité* regarding the remote Outer Hebridean Islands which comprise the conjoined Isle of Lewis and Harris and neighbouring islands of North Uist, Benbecula, South Uist and Barra. However, the peculiarities of this environment, perhaps the most distinctive of the UEP case studies, is reflected clearly within the perceptions and social practices of caring, stewardship and responsibility for parks and open spaces. Classified as predominantly rural, there is a population density of around 10 people per square kilometre and just 8500 living in the main town of Stornoway on the Isle of Lewis (Currie et al. 2019, p. 5). Historically, the patterns of land ownership and use are based on crofting, where small parcels of land were tenanted for subsistence along with access and use rights to common land located just outside settlements, for the benefit of the landowners who profited from the surplus labour of crofters forced to work in the fishing and seaweed kelping industries that they controlled. These patterns changed on Lewis when in 1923 landowner Lord Leverhulme, the Edwardian industrialist, transferred nearly 70,000 acres surrounding Stornoway to the community under management of the Stornoway Trust. This stimulated the beginning of community land buy-out and trust development across the Western Isles which continues today.

Scotland has the highest concentration of land ownership in the fewest hands, a legacy of the 'Clearances' of crofting communities 200 years ago. Nearly 83% of its land is in private hands, with the rest under control of the Forestry Commission, the Scottish government and devolved elements of the Crown Estate (Hetherington 2015, p. 82) but 50% of this land is owned by around 430 owners, with 10% owned by just 16 individuals or companies. Two-thirds of substantial rural land stock, 85% of which is unsuitable for cultivation, is owned by just 0.025% (Christophers 2019, p. 52). As Christophers suggests, the extreme concentration of private ownership with its potent for value extraction is not simply about wealth but about the deprivation of those to whom it is denied: 'the locus of power not just to make money, but to shape our collective societies, economies and environment' (p. 56). Ownership and access to land is, therefore, at the centre of debates about power and democracy in Scotland and the subject of successive attempts at land reform by the Scottish National Party government during the process of devolution and ongoing campaign for independence.

The participation in community land ownership in the Western Isles is both a radical new direction for Scotland and the vestige of past ways of planning and managing the environment. Alongside the residual practices of crofting which include keeping livestock, common grazing and occupational pluralism (Nascimento 2019), the islands retain widespread Gaelic language speaking. These practices shape the expectations and perceptions of those involved in place governance, the land trusts, local authorities, governmental agencies such as the Crofting Commission and the development agency, Highlands and Islands Enterprise, and inform the dynamics of place and sense of belonging of our research interviewees, which are also structured by status of islander, incomer, second householder and tourist. Senior (2018), whose year-long ethnography in the Western Isles alongside household interviews comprised the UEP case study research, observed that community land buyouts were motivated by the stakes of local people in revitalising the area for themselves and their neighbours, but also that there was intrinsic value attached to civic participation in everyday community life. She describes the 'super-participators' (p. 2), younger members of community land trusts who actively volunteer in many different additional activities and aspire to take on the responsibility of becoming trust directors in later life; these she finds may be benefited from growing up with the civic ideals of the islands, but despite this face resistance from older members of the community who do not take them seriously.

Amongst the interviewees, there are diverse opinions over the benefits of participation in community development and ambivalence over who should and could take part, and why. Some of the interviewees were indeed super-participators, who created plans for safe places for exercise on the moorland, for community halls, playpark equipment, and shipping containers to provide spaces for groups such as Man Sheds, motivated by the social purpose of participation. There were also some who were reluctant to take part in residents associations and community initiatives, as they wanted to participate for the intrinsic value of volunteering, not for the power or authority:

> One of the reasons I don't get involved is because I find the people who get involved are people who like to be in charge of these things, they love having authority and they're the people I don't really like […] And I just can't be doing with that, so I mean you're in it because you love it, not to, oh, let's see how much money we can get out of whoever, whatever body to fund this exhibition. (Stornaway participant)

Others actively remained passive, acknowledging their particular island status:

> I think as an incomer you don't want to get too involved with things like that […] there is somebody we know who I think has put a lot of backs up—, an incomer, put a lot of backs up by being very controversial and critical and I don't really think it's the done thing. That's not a role we want to play, we kind of go along quite quietly. (Stornaway participant)

Some were motivated by their frustration with the commercialisation of traditional spaces for communing, such as pubs and shops which had changed owners and use to become restaurants or high-end retail, something they saw as 'pricing out' of everyday participation. There was a concern about the privatisation of public space which had previously been provided by a community play park:

> 20 years ago they took a bit of land off the Trust to put in a park. And it went well for the first ten years but as that first generation of kids left, the interest went. And then the ones who came after, by that time—, this is early 2000s. By that time the explosion in finance had taken place in the town and everybody was then going in and they were going away on holidays, and they were, you know. They started putting the play parks in their own gardens, you know. The big swing sets and the trampolines and the—, so when you've got the people with money with the play parks in their own gardens they tend to lose interest in the communal play park, which is what happened. (Stornoway participant)

The same park was rumoured to be on valuable but polluted land, and there were suspicions that this too would be remediated then turned over into housing development, rather than return to its park status. Even for community land trusts, there was an underlying threat of development and disagreement over how to manage the land in ways to make the community economically sustainable whilst retaining its natural and cultural assets. The promotion and state subsidy of wind turbines, which provide renewal energy and additional income for community land trusts who had control over the use of their land even when they did not formally own it, was a further cause for concern for the damage it might cause to the natural beauty of area, seen as an important driver for tourism income. The case for a National Park designation which would have brought statutory protection of the landscape, led by the North Harris Land Trust, was rejected in 2010 due to a lack of evidence of economic benefits (Mackenzie 2012).

Balancing these conflicting concerns over sources of revenue and inward investment is a matter of negotiating local and national policy and planning landscapes, and takes skill, commitment and labour. Community trusts are becoming 'key local governance actors' (Currie et al. 2019, p. 10) who rely on opportunities for revenue such as renewable energy from wind turbines for their success and sustainability The opportunities for community land ownership and management, through crofting and land trusts, may not eradicate the tensions between economic sustainability, rights to public space and use of land; however, they provide means for their deliberation and negotiation. Over the last three decades, community-owned land has increased fivefold in Scotland, and the government has pledged to continue support, recognising its importance in reversing declining populations, revitalising localities and promoting sensitive sustainable development. In the Western Isles, nearly three-quarters of the population and two-thirds of crofts are situated on community-owned estates which cover more than half the landmass in the Western Isles. As Senior (2018) argues, suitable structures and policies are required to encourage the continued active participation of citizens and to sustain the common-pool resources that they govern.

Heading Back to the Original Modern City

Returning to Manchester-Salford, the location of those first municipal parks whose opening is discussed at the beginning of this chapter, the fieldwork included ethnography of local parks in North Manchester and East Salford, participatory workshops, household interviews and follow-on community engagement and practice research within three parks under the management of Manchester City Council: Cheetham Hill or Elizabeth Street Park, Whitworth Park and Platt Fields Park (also discussed in Chap. 3). As described in Chap. 4 and elsewhere (Gilmore 2017), the articulation of public value through everyday participation in the many public parks in the two neighbouring inner-city wards of Cheetham and Broughton was rich and insightful, confirming the broader value frameworks that are evidenced within policy literature, such as the benefits to health and well-being, social capital, land value and sense of place. The research also found evidence of the 'social enclosure' (Gilmore 2017) of parks through self-exclusion by different ethno-cultural groups whose beliefs and values meant they might avoid parks at certain times, or who were concerned about threats to their safety due to the participation of other groups, in particularly anti-social young people. There was also evidence of broader community safety concerns allied to the deterioration of the parks which also signalled a lack of care from those they deemed responsible for their upkeep. Accordingly, and resoundingly, the research found that interviewees had very strong expectations of who should care for parks, and therefore whom to blame for their decline, and that this was the local city council.

The community stewardship of public parks and open spaces was also the topic for debate in the context of austerity which palpably framed the UEP fieldwork. Meetings with our local authority research contacts were interrupted as phone calls asking for repairs to park toilets from officers whose duties of care stretched across libraries, museums, community centres, public realm, waste and parks management, or attendance cancelled due to lack of time, too many commitments and not enough staff. On one occasion, a meeting was postponed as the contact, a senior local authority officer, needed to find ways to cut their own budget, reportedly by over £1 million. Pilot projects to share services and good practice between the community libraries in neighbouring authorities were cancelled due to funding cuts, and directorates and teams were restructured to try to make service delivery more cost-effective.[4] The local authority officers were desperate, therefore, to encourage community participation in volunteering and stewarding of parks and open spaces which might fill the gaps within their capacity, through the creation of friends' groups and volunteer activities such as guerrilla gardening, community policing and litter-picking.

However, the demographics of the two areas were seen to impact community engagement in neighbourhood management negatively. Both areas are characterised by large ethnic minority populations and diverse religious faiths; this could lead to both homogeneity and heterogeneity in neighbourhoods, for example, the large Jewish community located in neighbouring wards, Kersall and Broughton, was estimated to be 32% of East Salford's population (Partners in Salford n.d., p. 6). Strategic priorities for both areas included addressing waste management, fly-tipping and providing access to opportunities for active participation to address significant health inequalities. There was the perception in local government that the ethnically and linguistically diverse and transient migrant populations of the Cheetham ward were wilfully engaging in cultural practices that created anti-social issues in the neighbourhood, whilst expecting service delivery. An example of this was when sports equipment for street cricket was given to the community, on the assumption that it would support youth engagement, but was used by taxi drivers in the local area gathered in the evening, to play cricket, socialise and gamble, leaving their rubbish behind and inciting the frustration of the local government officers. Another example was the practice of pigeon-feeding by South Asian communities in the parks and open spaces which was a form of recycling of food and connected to faith, but since it encouraged vermin, it was seen as a further sign of the lack of responsibility from resident communities over where they lived (Gilmore 2017).

In Broughton, East Salford, there was a short-lived attempt to encourage members of the orthodox Jewish community to take over a space in one neighbourhood park through funding to start up a community growing space that might become an income generator for the Salford parks team, or at least sustain a community space in one part of the park. The Council had no capacity to manage the project, however, or seemingly any understanding of the local community's interest in gardening or food cultivation, let alone the specific norms and practices of the community or the rhythms of their daily lives, which saw participation in parks change depending on the co-presence of gentil communities (Gilmore 2017). The lack of appreciation for cultural differences was also a factor in the failure to engender stewardship within Cheetham Park, during a follow-on project. In this we tried to engage local communities in 'commoning' by taking part in participatory activities in the park but did not take into account the different faith-based restrictions for participants on specific days of the week and religious holidays (Gilmore and Lang 2020).

During the UEP research, consultation was taking place for a new Parks Strategy for Manchester's 131 parks and 12 river valley green spaces which would replace the outdated strategy of 2001 and be aligned with the priority outcomes of a new community strategy for the city, Our Manchester Strategy, which set out a vision 'to put Manchester in the top-flight of world cities by 2025' (Manchester City Council 2016). Advocacy for this new strategy was made to a Neighbourhoods Scrutiny Committee early in 2016, setting out the need to address the decline of Manchester's green assets which 'have not always been maintained or achieve the standard a world class city would aspire to' (Manchester City Council 2016, p. 3). Setting out the claim for protecting the many environmental, health, recreational and social benefits for Manchester's residents, the advocates also put forward the economic case, citing the Heritage Lottery Fund State of UK Public Parks 2014: Renaissance to Risk report (HLF 2014), calling on the private sector as potential investors and citing parks' positive impact on property and land value as 'positive value-added investments, and a platform for "good economics" ' (Manchester City Council 2016, p. 4).

The final strategy was eventually adopted and launched in December 2017 (Manchester City Council 2017, 2021). Like Peterborough, it included a segmentation model which differentiated between 'Local parks': small spaces that provide a focal point for local neighbourhoods and include one or more of the following: a grass area, play area, sports facility, flowerbed; 'Community parks': larger local parks that have a variety of facilities and features and can host activities and small community events; and 'Destination parks', which include a variety of amenities, visitor attractions, distinctive features and open space that can be used in different ways, such as the hosting of major events (Manchester City Council 2017, p. 10). It established thematic priorities which brought together the physical role of parks within neighbourhoods, the forms of use and participation that they could provide, including events that could generate additional income, the development of a 'Manchester Quality Standard' to assess management and maintenance and deliver 'world class visitor destinations', and finally the theme of 'Productive Parks in Partnership ... to deliver park services in a more collaborative and fruitful manner with communities and local organisations, not just the Council' (Manchester City Council 2017, p. 16) which include the mandate for all parks to coproduce an action plan with communities (see Fig. 5.4).

Progress reports to internal scrutiny committees one year later reveal the coordination issues that the local authority faced, as it attempted to take forward its action plan. The first was in communicating with the many different groups involved in park stewardship, which detailed over 50 formal friends' groups alongside the many other geographically dispersed stakeholder groups who wanted to be involved in planning for their local parks. The strategy calls for the creation of local park plans involving multiple agencies which led to far more interest and desire for involvement than the parks team could support. Relatedly, the second was the culture of the parks team, which like other areas within the public sector workforce retained a large proportion who had been within their current positions for over 10 years, with limited opportunities for progression. The job of park manager was seen to be far more than one of park-keeping, more akin to a cultural manager or creative producer, given the need to fundraise, undertake community engagement and manage community festivals and events. An engagement programme was put in place to advise staff on how their work delivered the new strategy, whilst restructuring of outdated processes and roles began. The third was the ongoing decline in central government funding and the need to prioritise investment to those parks which could 'close the gap between ongoing trading income and expenditure' (Manchester City Council 2018) (see Fig. 5.5).

The council progress report following the outbreak of COVID-19 (Manchester City Council 2021) reflected on the difficulties caused by the pandemic, which saw a 30% increase in park visitors at times when toilets and playground facilities were locked shut due to social distancing measures when there were significant concerns about public health for any kind of gathering. It confirmed the City Council's ongoing commitment to investing in its parks, including a £960k Parks in Partnership Fund

Fig. 5.4 Manchester Parks Strategy and the mandate for participation in Park Action Planning. (Manchester City Council 2017, p. 13)

which provides seed funding for local park community projects, and to increasing income from charging with an expanded programme of events, such as the Mela, the Caribbean Carnival, large-scale music concerts and immersive arts experiences. Until the pandemic a key source of income

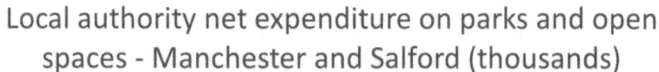

Local authority net expenditure on parks and open
spaces - Manchester and Salford (thousands)

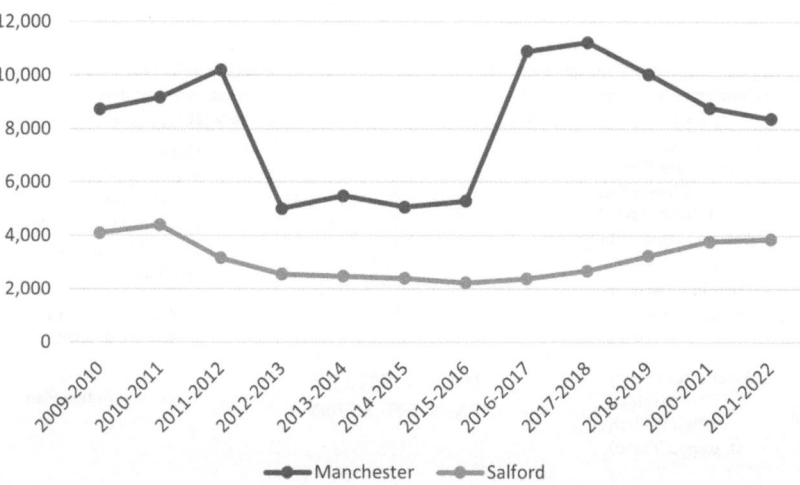

Fig. 5.5 Local authority net expenditure* on parks and open spaces, Manchester and Salford (thousands). (Source: DLUHC/MHCLG 2022) *Net expenditure comprises running and employee costs less income from sales, fees and charges, and grants passed on to third parties

was generated by the Parklife festival, a two-day music festival established in 2010 in Platt Fields Park before moving to Heaton Park to the north of the city. Parklife was established by local entrepreneur, Sacha Lord, who also founded the Warehouse Project, a highly successful brand of club nights in underground venues in Manchester, and who was invited in 2018 to be the city-regional Night-time Economy Adviser by the Greater Manchester Mayor Andy Burnham. The 80,000-capacity festival brings in an estimated £13 m to the local economy, supporting income for the parks team and a dedicated community fund in Manchester, Bury and Rochdale (Hookings 2019).

Such programming in public parks has been criticised in the work of Smith (2019) who identifies the problems of exclusion, enclosure and commodification that commercial events can bring. Looking at the London parks, Smith shows how some have seen frequent and long-term disruption from large-scale concerts and festivals, fenced off from local communities and against their wishes, whilst, he argues, the income does not necessarily not go back to local communities or the parks themselves, and may threaten the precarious new governance arrangements and small-scale innovations in business models recommended by national government (Smith 2019). The disruptions to local communities by concert goers making their way back into the city at night from Heaton Park was noted by local authority contacts within the UEP research; however, the importance of Parklife as a source of additional income was seen to override this temporary disturbance. The programming of Manchester bands such as Oasis and New Order and global rave artists alongside community festivals is homologous with the history of music programming in the city's parks (Gilmore and Doyle 2019). Bandstand concerts curated by the Parks Committee to educate park-goers in their music tastes also began to encourage more popular repertoire and to install bandstand enclosures and create ticketed areas to drive up income for 'playgrounds that pay for themselves' (p. 142). It is also homologous with Manchester's brand commissioned by the city council and created by New Order and Factory Records in-house designer, Peter Saville, in 2009, coinciding with the first edition of the Manchester International Festival. This proclaims Manchester as the 'Original Modern' city which claims its status as birthplace for the industrial revolution as the source of inspiration for ongoing creative production, scientific discovery and civic innovation (Marketing Manchester 2009).

CONCLUSION

Trying to encapsulate the 'local structures of feeling' (Taylor et al. 1996) of the places described above in this short analysis feels like the Original Modern brand's attempt to convey the key messages of Manchester as a radical 'creative city' (Whiting et al. 2022): condensed and somewhat self-conscious. However, the case studies above provide hints of how participation narratives (Miles 2016) articulate not only the values of everyday participation in parks but also the policy thinking, statutory frameworks, funding models, regulatory technocracy and that surround parks, their communities of interest and their future sustainability in England and Scotland.

This chapter has moved beyond looking at the logics and meanings of participation in parks to examine participation in their stewardship as a form of governance of place. I have examined the responsibilities of local government as the principal owner-managers of parks and argue that these are locked into a subordinate relationship with the central state, which in England has refused the means for sufficient resourcing from public funds and lacked the will to support land reform or statutory mechanisms to protect municipal parks despite significant evidence of their public value and benefit. This is a different case in Scotland where the discourses of independence and the emotional historical baggage of the clearances have complicated capitulation to the overwhelming unequal state of land ownership which, combined with particular needs of rural remote populations, have ushered in more progressive opportunities for community land ownership and local statutory duty to ensure equality of access to parks and open spaces.

The strength of public sentiment for parks means that local authorities are under pressure to sustain these spaces, which are also rivalled and depleted by their use. It reveals another paradox: in order to function as publicly valuable common-pool resources, parks require the participation and collaboration of their publics to act together in governance, to maintain social order and ensure that the resources are not rivalled or depleted. Yet, making 'productive parks in partnership' requires resources not solely for upkeep or new capital investment but for managing change, engaging communities, for integrating policy thinking and planning. However, the alternatives to state funding such as community transfer, land trust management and new business models challenge the capacity of local communities and local authorities alike and have ramifications for democracy.

Examples from the UEP project consider the affective relations of place governance in diverse ecosystem case studies which are revealed in the perceptions of local residents and strategies of local authorities for caring for their public parks. These I argue show the situated practices of policy and its grounding in 'local structures of feeling' (Taylor et al. 1996) reflected in the properties and dynamics of place, or following Sennett (2018), the *ville* and the *cité*. This is not to argue there is structural determinism within place-type or that it is enforced top-down by local governance, hardly the case when local government is so constrained and disempowered to solve the park value paradox. Rather it is to reinforce the need to examine local affective relations and address state responsibilities, both national and local, in supporting cultural democracy through the provision of public parks as spaces for everyday participation and practising citizenship.

In the following concluding chapter, I bring together the various empirical observations and theoretical propositions for practising sustainable democratic public space within the preceding chapters to consider further the question of park equity and the right to the park as the right to the city (Harvey 2003, 2020), a fundamental feature of cultural democracy.

NOTES

1. As outlined in Chap. 1, this mixed methods research project was funded by the Arts and Humanities Research Council Connected Communities 'Communities, Culture and Creative Economies' programme (AH/J005401/1) with partnership funding from Creative Scotland. For further information see www.everydayparticipation.org
2. Heritage Lottery Fund became the National Lottery Heritage Fund in a rebranding exercise in 2019.
3. There are 46 Areas of Outstanding Natural Beauty (AONB) in England, Wales and Northern Ireland, 34 of which are in England. Designated by Natural England, long with National Parks, all have statutory protection under the Countryside and Rights of Way Act 2000 (CROW Act), although only National Parks have Planning Authority (Natural England 2018). Three of the AONBs in Devon are wholly within the County (North Devon Coasts, East Devon and South Devon) and two (Tamar Valley and Blackdown Hills) are cross-boundary with neighbouring authorities. Taken with Dartmoor and Exmoor National Parks, these 'protected landscapes' cover 35% of Devon (Devon County Council n.d.).
4. Manchester City Council undertook a major restructure in 2015, following a decrease of £56 m or nearly 16%, in its central government settlement from the previous year. This represented a cut of 9.4% in the Council's Spending Power, the measure of revenue finance available to local authorities, which, compared with the national average of 1.8%, was one of the highest cuts to local government spending in the country. The strategic response from the council was to restructure services with a focus on neighbourhoods, with citywide service areas, including 'Libraries, Galleries and Culture' delivering to integrated neighbourhood teams. At the same time, further investment was announced in capital projects related to Heaton Hall and Heaton Park, and to underwrite increase budget for the new cultural facility in the city centre, Home, which was funded by external contribution (Manchester City Council 2015).

REFERENCES

Barker, A., A. Crawford, N. Booth, and D. Churchill. 2019. Everyday encounters with difference in urban parks: Forging 'openness to otherness' in segmenting cities. *International Journal of Law in Context* 15 (4): 495–514. https://doi.org/10.1017/S1744552319000387.

Bramhill, P., and D. Tibbatts. 2002. *Public park assessment: A survey of local authority owned parks focusing on parks of historic interest.* Urban Parks Forum.

Bryson, J.M., B.C. Crosby, and L. Bloomberg. 2014. Public value governance: Moving beyond traditional public administration and the new public management. *Public Administration Review* 74 (4): 445–456. https://doi.org/10.1111/puar.12238.

CABESpace. 2009. *Making the invisible visible: The real value of park assets* [online]. http://webarchive.nationalarchives.gov.uk/20110118104354/http://www.cabe.org.uk/publications/making-the-invisible-visible. Accessed 03 Apr 2023.

Cambridgeshire County Council. n.d. Cambridgeshire open spaces mapping and strategy summary report. https://cambsfutureparks.org.uk/wp-content/uploads/2021/06/cambridgeshire-open-space-mapping-and-standards.pdf. Accessed 10 Apr 2023.

Cambridge Open Space. n.d. About Us https://www.cambsopenspace.co.uk/about Accessed 22 Dec 2023.

Christophers, B. 2019. *The new enclosure: The appropriation of public land in neoliberal Britain.* London: Verso.

CLG. 2017. Communities and local government committee public parks inquiry. https://publications.parliament.uk/pa/cm201617/cmselect/cmcomloc/45/45.pdf. Accessed 19 Feb 2022.

Conway, H. 1991. *People's parks. The design and development of Victorian parks in Britain.* Cambridge: Cambridge University Press.

Conway, H., and D. Lambert. 1993. *Public Prospect: Historic urban parks under threat: A short report by the Garden History Society and the Victorian Society.* London.

Crowe, L. 2018. The future of public parks in England: Policy tensions in funding, management and governance. *People, Place and Policy* 12 (2): 58–71.

Currie, M., A. Pinker, and A. Copus. 2019. *Strengthening communities on the Isle of Lewis in the Western Isles, United Kingdom,* RELOCAL case study N° 33/33. Joensuu: University of Eastern Finland.

Devon County Council. n.d. National and UNESCO designations. https://www.devon.gov.uk/environment/landscape/national-and-unesco-designations. Accessed 12 Apr 2023.

Di Masso, A. 2015. Micropolitics of public space: On the contested limits of citizenship as a locational practice. *Journal of Social and Political Psychology* 3 (2): 63–83.

Dickinson, J., E. Bennett, and M. James. 2019. Challenges facing green space: Is statute the answer? *Journal of Place Management and Development* 12 (1): 121–138. https://doi.org/10.1108/JPMD-09-2017-0091.

DLUHC/MHCLG. 2022. Local authority revenue expenditure and financing. Department for Levelling Up, Housing and Communities and Ministry of Housing, Communities & Local Government. https://www.gov.uk/government/collections/local-authority-revenue-expenditure-and-financing. Accessed 10 May 2023.

Durham, M. 1995. Lottery cash to fund city parks clean-up. *The Observer*, May 21, p. 3.

Gateshead City Council. n.d. Saltwell Park https://www.gateshead.gov.uk/article/3958/Saltwell-Park. Accessed 22 Dec 2023.

Gateshead Council. 2013. Green Gateshead: Green infrastructure delivery plan 2013–2030, p1–35. https://www.gateshead.gov.uk/media/3844/Green-Infrastructure-Delivery-Plan-2013-2030/pdf/GreenInfrastructureDelivery DeliveryPlanJanuary2013_1.pdf?m=636443562438130000. Accessed 10 Apr 2023.

Gibson, L. 2019. Cultural ecologies: Policy, participation and practices. In *Histories of cultural participation, values and governance. New directions in cultural policy research*, ed. Eleonora Belfiore and Lisanne Gibson, 153–182. London: Palgrave Macmillan.

Gibson, L., M. Taylor, and D. Edwards. 2014. Understanding everyday participation in Gateshead briefing document, July 2014, published by the AHRC Understanding Everyday Participation project. www.everydayparticipation.org. Accessed 12 Jul 2022.

Gilmore, A. 2017. The park and the commons: Vernacular spaces for everyday participation and cultural value. *Cultural Trends.* https://doi.org/10.1080/09548963.2017.1274358.

Gilmore, A., and P. Doyle. 2019. Histories of public parks in Manchester and Salford and their role in cultural policies for everyday participation. In *Histories of cultural participation, values and governance. New directions in cultural policy research*, ed. Eleonora Belfiore and Lisanne Gibson, 129–152. London: Palgrave Macmillan.

Gilmore, A., and L. Lang. 2020. Talking, walking and making in Cheetham Park. *Conjunctions* 7 (2). 1- 20. ISSN 2246-3755.

Greenhalgh, L., and K. Worpole. 1995. *Park life.* Stroud/London: Comedia and Demos.

Harding, S. 1999. Towards a renaissance in urban parks. *Cultural Trends* 9 (35): 1–20.

Harvey, D. 2003. The right to the city. *International Journal of Urban and Regional Research* 27 (4): 939–941.

———. 2020. The right to the city: New left review (2008). In *The city reader*, ed. Richard T. LeGates and Frederic Stout. 281–289. London: Routledge.

Heritage Lottery Fund. 2014. *State of UK public parks: Renaissance to risk.* London: Heritage Lottery Fund.

———. 2016. *The state of UK public parks.* London: Heritage Lottery Fund.

Hetherington, P. 2015. *Whose land is our land?* Bristol: Policy Press.

Historic England. 1985. Saltwell Park: Official list entry. https://historicengland. org.uk/listing/the-list/list-entry/1001182?section=official-list-entry. Accessed 25 Apr 2023.

Hookings, M. 2019. Parklife festival generates £13m worth of economic benefit for Greater Manchester. *Event Industry News.* https://www.eventindustrynews.com/news/parklife-festival-generates-13m-worth-of-economic-benefit-for-greater-manchester/. Accessed 16 Apr 2023.

Kelsey, T., and M. Kenny. 2021. *Townscapes: The value of social infrastructure,* 1–65. Cambridge: The Bennett Institute for Public Policy, University of Cambridge.

Layton-Jones, K. 2016. History of public park funding and management (1820–2010). Historic England London, 135.

Lee, A., H. Jordan, and J. Horsley. 2015. Value of urban green spaces in promoting healthy living and wellbeing: Prospects for planning. *Risk Management and Healthcare Policy* 8: 131–137.

MacKenzie, A. Fiona D. 2012. *Places of possibility: Property, nature and community land ownership.* Wiley.

Manchester City Council. 2015. Manchester City Council report for resolution to: Neighbourhoods Scrutiny Committee – 10 Feb 2016, Communities Scrutiny Committee – 11 February 2015; Economy Scrutiny Committee – 11 February 2015; Executive – 13 February 2015; Finance Scrutiny Committee – 23 February 2015; Subject: Budget Proposals for Growth and Neighbourhoods 2015-17 report of: Deputy Chief Executive (Growth and Neighbourhoods) and City Treasurer, Item 7 Manchester Executive, 13 February 2015.

———. 2016. Manchester City Council report for information report to: Neighbourhoods Scrutiny Committee – 23 Feb 2016, Subject: Update on parks strategy report of: Deputy Chief Executive (Neighbourhoods), Strategic Lead (Parks, leisure & events) Item 7 Neighbourhoods Scrutiny Committee 23 Feb 2016.

———. 2017. Manchester City Council report for resolution.

———. 2018. Manchester City Council report for information report to: Communities and Equalities Scrutiny Committee – 24 May 2018, Subject: Manchester's park strategy 2017–2026 report of: Deputy Chief Executive (Growth & Neighbourhoods), Director of Neighbourhoods, Strategic Lead (Parks, Leisure and Events).

————. 2021. Manchester City Council report for information report to: Environment and Climate Change Scrutiny Committee – 24 June 2021; Subject: Manchester's park strategy – Progress through the pandemic report of: Strategic Director (Neighbourhoods).

Marketing Manchester. 2009. Original modern, first published February 2009. http://www.marketingmanchester.com/wp-content/uploads/2017/02/OriginalModern.pdf. Accessed 28 Apr 2023.

Martinsson, K., D. Gayle, and N. McIntyre. 2022. Funding for England's parks down £330m a year in real terms since 2010. *The Guardian*, August 23. https://www.theguardian.com/environment/2022/aug/23/funding-for-englands-parks-down-330m-a-year-in-real-terms-since-2010. Accessed 10 May 2023.

Miles, A. 2016. Telling tales of participation: Exploring the interplay of time and territory in cultural boundary work using participation narratives. *Cultural Trends* 25 (3): 182–193. https://doi-org.manchester.idm.oclc.org/10.1080/09548963.2016.1204046.

Miles, A., and J. Ebrey. 2017. The village in the city: Participation and cultural value on the urban periphery. *Cultural Trends* 26 (1): 58–69. https://doi.org/10.1080/09548963.2017.1274360.

Nascimento, J. 2019. Working the fabric: Resourcefulness, belonging and Island life in the Harris Tweed industry of the Outer Hebrides of Scotland, A thesis submitted to the University of Manchester for the degree of Doctor of Philosophy in the Faculty of Humanities.

Natural England. 2018. Areas of outstanding natural beauty (AONBs): Designation and management. https://www.gov.uk/guidance/areas-of-outstanding-natural-beauty-aonbs-designation-and-management. Accessed 10 May 2023.

NESTA. 2018. *Meet the rethinking parks innovators: Eight parks projects developing promising and innovative operating models.* London: NESTA. https://www.nesta.org.uk/blog/meet-rethinking-parks-innovators/. Accessed 30 Oct 2018.

NLHF. n.d. Public parks and urban green spaces. https://www.heritagefund.org.uk/our-work/landscapes-parks-nature/public-parks-urban-green-spaces. Accessed 03 Apr 2023.

Office of the Deputy Prime Minister. 2003. Housing, planning, local government and the regions committee, Eleventh report of session 2002–03, Living places: Cleaner, safer, greener, HC 673-I. https://publications.parliament.uk/pa/cm200203/cmselect/cmodpm/673/673.pdf. Accessed 05 Apr 2023.

Partners in Salford. n.d. East Salford neighbourhood baseline profile.

Rex, B., and P. Campbell. 2022. The impact of austerity measures on local government funding for culture in England. *Cultural Trends* 31 (1): 23–46. https://doi.org/10.1080/09548963.2021.1915096.

Ruff, A. 2016. *Manchester's Philip's Park: A park for the people by the people since 1845.* Stroud: Amberley Publishing.

Ryan, F. 2017. The uncomfortable truth: UK government cuts have happened under our noses. *The Guardian*, February 28. www.theguardian.com/public-leaders-network/2017/feb/28/UK-government-cuts-parks-libraries-local-government-nhs-prisons. Accessed 12 Apr 2023.

Schaefer, K., D. Edwards, and J. Milling. 2017. Performing Moretonhampstead: Rurality, participation and cultural value. *Cultural Trends* 26 (1): 47–57. https://doi-org.manchester.idm.oclc.org/10.1080/0954896 3.2017.1274359.

Senior, L. 2018. Young adults and community land management, Understanding Everyday Participation Research Briefing, July 2018. www.everydayparticipation.org. Accessed 10 May 2023.

Sennett, R. 2018. *Building and dwelling: Ethics for the city*. London: Allen Lane.

Smith, D.A. 2019. The tyranny of the temporary: Should we be worried about the rise of park events? Historic England. https://historicengland.org.UK/whats-new/debate/recent/public-parks/tyranny-of-the-temporary/. Accessed 20 Nov 2019.

Tandy, V. 2019. The Heritage Lottery Fund and its role in the construction and preservation of the past: 1994–2016. PhD thesis, University of Manchester.

Taylor, I.R., K. Evans, and P. Fraser. 1996. *A tale of two cities: Global change, local feeling, and everyday life in the north of England: A study in Manchester and Sheffield*. London: Routledge.

The City of Peterborough. 2019a. Assessment of parks and open spaces executive summary Appendix A. https://www.peterborough.ca/en/city-hall/resources/Documents/Studies-and-Projects/Parks-and-Open-Spaces/CSRS20-003%2D%2D-Appendix-A%2D%2D-Assessment-of-Parks-and-Open-Space-Exec-Summary.pdf%20. Accessed 10 Apr 2023.

———. 2019b. Parks development standards Appendix D. https://www.peterborough.ca/en/city-hall/resources/Documents/Studies-and-Projects/Parks-and-Open-Spaces/CSRS20-003%2D%2D-Appendix-D%2D%2DPark-Development-Standards.pdf. Accessed 10 Apr 2023.

The Parks Alliance. n.d. Why parks matter: An evidence base. https://www.theparksalliance.org/why-parks-matter-evidence-base/. Accessed 09 Apr 2023.

The Parks Trust. n.d. Our work, our story. https://www.theparkstrust.com/our-work/our-story/. Accessed 04 Apr 2023.

Urban Green Newcastle. n.d. About Us https://urbangreennewcastle.org/ Accessed 22 Dec 2023.

Walmsley, B., A. Gilmore, D. O'Brien, and A. Torrigiani. 2022. *Culture in crisis: Impacts of Covid-19 on the UK cultural sector and where we go from here*. Leeds: Centre for Cultural Value.

Wells, P. 2018. Evidence based policy making in an age of austerity. *People, Place and Policy* 11 (3): 175–183.

Whiting, S., T. Barnett, and J. O'Connor. 2022. 'Creative city' R.I.P.? *M/C Journal* 25 (3). https://doi.org/10.5204/mcj.2901.

CHAPTER 6

The Political Economy of Contemporary Public Parks

Abstract Following a review of the arguments presented in previous chapters, this concluding chapter returns to consider parks, private interest and public good in the contemporary city, through the examples of two city centre parks in Manchester, made 100 years apart. Building on the history of park-making in the nineteenth century as a form of municipalisation, the chapter examines the context where a combination of outsourcing and privatisation of public park maintenance, local authority budget cuts and the encroachment of private interest in realising land value is resulting in the de-municipalisation of place governance. It considers how these processes prevail on the right to the city, following Lefebvre (Le Droit à la ville. Anthropos, Paris, 1968), which is manifest in the public park not just as a text on the city but as a right to everyday participation in urban life and cultural democracy.

Keywords Municipalisation • De-municipalisation • Right to the city • Right to the park • Place governance • Privately owned public space (POPS) • Cultural democracy • Outsourcing • Privatisation

© The Author(s), under exclusive license to Springer Nature Switzerland AG 2023
A. Gilmore, *Culture, Participation and Policy in the Municipal Public Park*, Palgrave Studies in Cultural Participation, https://doi.org/10.1007/978-3-031-44277-3_6

Introduction

A central argument of this book is that we should consider municipal public parks as part of the broader cultural policy infrastructure that shapes and governs places and people. I have been considering this question by examining how people value their everyday cultural lives and the meanings and feelings they attach to their practices of everyday participation. In turn, I argue that this participation contributes to the cultural management of places through compliance with and resistance to policy and planning and through what Di Masso (2015) calls the emplaced practices of citizenship. I have also been concerned with the structural and spatialised power relations that are found within the discourses and practices of those who make, manage and advocate for public parks, the rationales and philosophies they espouse and what these tell us about the relationships between the civic society and the state, both central and municipal, which manifest as exchanges and withholding of public and private good. To do so, I have called on key concepts which are outlined in Chap. 1 and built on in the following chapters, to articulate a theoretical framework which links cultural policy, participation and place. These concepts are used to critically evaluate empirical data from the Understanding Everyday Participation (UEP) project, in conjunction with follow-on qualitative interviews, archival research and policy analysis. The case studies of parks and museums in English and Scottish places have a particular focus on the industrial towns and cities in northern England of Manchester, Macclesfield and Sheffield, contextualised with reference to further examples from the United States and Europe. Though based on research in the north-western hemisphere, the themes and findings are arguably universally applicable since they contribute to an understanding of the state/cultural relations of everyday participation, cultural policy and place that is not confined to particular territories, hierarchies of scale or indeed historical epochs. In this final chapter, I want to explore the current dilemmas for municipal public parks, to pose some questions for their future and to consider the significance of these to cultural policy and place governance. Ahead of this discussion, it is useful to briefly recap arguments made in previous chapters.

Chapter 1 introduced the key concepts through which I consider everyday participation in municipal public parks and the cultural values derived through this participation. The discussion of cultural value has been a centrifugal point for cultural policy studies in the last two decades especially to policy scholars in the Global North whose research has a close relationship with and interest in the rationales for public support for arts and culture.

Researchers have explored ways to identify, measure and extrapolate cultural value, as part of taking cultural policy as object of study in its own right (Bell and Oakley 2014), but also in the pursuit of understanding how public policy might be influenced through research which corroborates 'evidence' of cultural value to underpin decision-making (Belfiore 2022). Like the term 'culture' (Williams 1983), cultural value is tricky, slippery and difficult to isolate, and the corroboration inevitably takes the form of quantifiable values through which to articulate economic, social, behavioural, aesthetic and physical effects connected with arts, cultural and creative production and consumption, that might be attached to government agendas (Gray 2002) as an instrument for their delivery (Belfiore 2020).

Even when conscious of this complicity, research on cultural value has become part of this question for 'evidence'. However, despite the maturity of cultural policy studies, its long tussles with ethical questions of the distinctions between research, evidence and advocacy, and entire research programmes dedicated to cultural value (Kaszynska and Crossick 2016), there has been (thankfully) no single accepted formula for its calculation which may serve to underpin such economic rationalism. Instead, as Belfiore urges, there is a need to roll-back from instrumentalism within cultural policy studies research, and to address 'a different reality from the ideal (and idealized) scenario of evidence-based policymaking' (Belfiore 2022, p. 300). Such a reality might be revealed by understanding policy processes as 'bounded rationality' (Cairney 2016, p. 5), tempered by political interest and expediency, which avoids the problem of the need for rigorous evidence and of competence on behalf of policymakers to interrogate and accept this evidence.

I have been careful, therefore, to make conceptual choices that can critically address this bounded rationality to reveal what and who constructs its boundaries and thereby controls access to the resources for their maintenance and erosion. Rather than focus on cultural value, I consider public space as something that is practised culturally, through gesture, social encounter, interaction, mediation and participation, and that this participation contributes more broadly to cultural policies and place governance. Key concepts include mediated conviviality (Barker 2017), locational citizenship and emplaced practise (Di Masso 2015), the cultural public sphere (McGuigan 2004), open forms, the *cité* and the *ville* (Sennett 2018), the commons (Ostrom 1990; Standing 2019) and the undercommons (Harney and Moten 2013), as discussed further below.

Chapter 2 concerned the history of municipal public parks in England and provides examples of the policy rationales and actors that were instrumental in their establishment. I argue that there is a symbiosis between the political economy of park-making and broader strategies of cultural management, the development of civic institutions and of the management of the city as a public body. Along with other new civic and cultural institutions, the making of public parks in the nineteenth century was not only an important strategy for social and moral improvement, managing public space and urban populations, but also a fundamental form of municipalisation which saw the responsibility of the public park become that of local government. I identify the 'public park value paradox' whereby through the development of these spaces for public policy the resulting business models ultimately challenge their own sustainability, especially in the context of state retraction, neo-liberalism and de-municipalisation, as is discussed further in Chap. 5.

In Chap. 3, I looked at museum-in-park-making, examining case studies of museums co-located in parks in Manchester, Macclesfield and Sheffield and the stories of the 'art reforming' advocates who made them, whose affective communications and philanthropic ambitions constituted the cultural public sphere of these places, and were responsible for creating and curating public art galleries and museums. The discursive practices of municipal museum-making in nineteenth-century northern England share similar policy rationale to those of park-making, concerning moral improvement, social mixing, education in arts, culture and science, and the means to regulate the public body. There are distinctions, however, particularly in the civic buildings and the challenges they present in meeting these policy aims which require those publics to move out of the park and over the threshold (Gurian 2005). The chapter examines contemporary examples of museums in parks and the strategies they employ to create the literate publics to enter into them, now understood as community engagement, audience development or socially engaged practice. The chapter argues that the open forms and opportunities for serendipity and synchronicity found in parks are more generative of public spaces that support cultural and counter-public spheres, than those of the buildings that sit within them, despite the best intentions of their creators, curators and directors.

In Chap. 4, I considered how participation practices provide fundamental influences which shape people's lives and everyday experiences, founded in their public parks but carried into broader relations of space and place through the social values, biographical narratives and meanings that become attached through practices of everyday participation. The chapter contrasts the different ecosystem case studies in the UEP project, exploring how people articulate and navigate their relationships with public space, and within their broader environments and senses of place. These meanings and values are multiple, multiperspectival and overlain; they overlap the experiences of the research participants, and present symbolic and generative values within their lives, as spaces for social inclusion and the formation of friendship and memories, and waymarkers in their life courses and their neighbourhoods. The municipal public parks and open green spaces of the case studies are places which can be enclosed through boundaries of ethnic, cultural and religious practice and social status, but also ones in which other constituencies' interests are visible and mediated, providing lower thresholds for 'mediated conviviality' (Barker 2017), for side-by-side living which may involve conflict and resistance as well as co-existence. As spaces for practising citizenship and public-ness, they also offer the opportunities for commoning, for community stewardship and cultural democracy, and for the articulation of cultural values that resist economic rationality or easy quantification.

I return to the cultural ecosystems of UEP project in Chap. 5, to consider what the participation narratives from this research propose in answer to the question of who should care for and steward public parks. These are considered in the context of legislative and statutory instruments for the municipal management of parks and green spaces in England and Scotland, which reflect the paradox of public value identified earlier in Chap. 1. Parliamentary enquiries, research and advocacy reports and media articles present evidence of the multifarious extrinsic values that parks offer and their overwhelming popularity with publics. However, their publicness and value depend on the ownership and maintenance by local authorities who are undergoing a sustained period of 'creeping neo-liberalism' (Crowe 2018) and swingeing budget cuts from central government. Hence the paradox: to function as common-pool resources, parks require the participation and collaboration of their publics to act together in governance, to maintain social order and lobby for resources. The more popular parks are the more these resources may be challenged, rivalled or depleted. There are models for commoning, such as community transfer and community land trusts; however, these require investment in engagement, complex regulation and integrated policy thinking and planning. With an absence of central government leadership or the policy instruments to ensure equal rights and access to land, the recommendations of parliamentary committees have stopped short of statutory regulation of the parks sector, and instead encouraged piecemeal approaches to pilot creative and innovative business models as alternatives to state funding.

Below, in this chapter, I continue to explore how thinking about public parks as cultural policy spaces exposes the challenges for their management and for broader governance of place, and revisit the idea of the urban commons, the relations between civic society and the state, and between public good and private interest. Firstly, the chapter examines the increasing threats to equitable access to parks in England, the rise of neoliberalism, the decline of municipal power and their impacts on park and green space equity. Secondly, I go back once more to Manchester to examine its newest park, Mayfield Park, in comparison to its most recent predecessor, Piccadilly Gardens, to draw out and link arguments concerning the right to the park as the right to the city (Harvey 2003, 2020) and the future public park as cultural public sphere. I conclude the chapter, and the book, by reviewing how this contributes to our understanding of cultural value, policy and place governance.

DE-MUNICIPALISATION AND THE ENTREPRENEURIAL PARK

On 23 April 2023, left-wing UK broadsheet *The Guardian* published an editorial on the newspaper's view on the public park, subtitled 'an asset that should be for everyone' (The Guardian 2023). This was far from the first such article, and its substance repeats concerns made in previous articles in the same left-wing broadsheet (e.g. Martinsson et al. 2022; Horton 2022) about the commodification and privatisation of public parks through the continuing strictures of austerity for local government in England, leading to an increasing inequity of access to parks and green spaces for urban populations. The newspaper editorial sets out evidence of the decrease in local government expenditure on public parks over the last decade due to the central government funding cuts of the post-crash austerity agenda, following the advent of a Conservative-led coalition government in 2010. In the decade following, the central government grants to local authorities decreased in real terms by 37%, reducing local authorities' spending power by 16% (Institute for Government 2020). As non-statutory services predominantly owned and managed by local government, the impact on parks spending by local authorities has amounted £327m in real terms between 2010–2011 and 2021–2022 with vast predicted costs increases due to energy price inflation and rising cost of living to come (Martinsson et al. 2022). *The Guardian* researchers identify the difficulties caused for local authorities by their strategies for raising income to 'plug the gaps' as park spaces are temporarily privatised through their rental as commercial festival and event spaces, which prevent local community use and nuisance. Furthermore, this 'thinning' of municipal infrastructure is overlaid on the privatisation and closure of other civic spaces, such as leisure centres and libraries, and unevenly distributed exacerbating inequality, as 87% of local authorities within the lowest quintile cut their spending compared to 58% of those in the most affluent (Martinsson et al. 2022).

The commodification of park spaces is a focus for Smith in his work on the leasing of London parks for festivals and events and the commercialisation of their amenities, examined within the broader context of neoliberal austerity and an enforced sharing of management beyond the state (Smith 2019, 2021). Those who live in the vicinity of parks suitable for large live events are subject to noise, traffic controls, anti-social behaviour, littering and the general disruption of having large-scale events and their audiences turn up in the neighbourhood. Furthermore, they are literally fenced out of their everyday participation. There are some positive effects of the disputes within stakeholder constituencies that arise from these temporary enclosures, such as the opportunity for collectivism and to recognise common interests across normally private concerns; however, these are limited. Furthermore, as Smith argues, this is an ideological move by local government to commercialise parks spaces and remove their responsibility from the state; whilst motivated by revenue extraction, this is part of a wider process of the commodification of the city and its assets (Smith 2019, 2021).

The increasing commercialisation of parks space is part of the experimentation with parks management heralded by the Community and Local Government Select Committee (CLG 2017) piloted by Heritage Lottery Fund and Nesta partnership projects to 'rethink' and encourage creative innovation, social enterprise and different governance models, as discussed in Chap. 5. This re-evaluation of public parks as commodifiable reveals how contradictory values can be rationalised within policy thinking simultaneously: the actual asset register of this public land value may well be low (park land is historically often gifted or bought at discount and cannot be sold on without significant political damage), whilst public values are judged as high according to a variety of econometrics. Commercialisation understands or 'rethinks' the value of the park space as a commodity, and thus temporary capitalisable private good.

Smith et al. (2023) identify these processes as a form of austerity-led 'de-municipalisation' (p. 2), emerging as the vestiges of new public management, privatisation, and emphasis on social entrepreneurialism and innovation, in lieu of public subsidy, investment and regulation. The experimentation in income generation, governance models and asset transfer is the result of *both* local authority budget cutting and the refusal of the national government to consider making parks provision a statutory duty, as outlined in Chap. 5. It amounts to, they argue, a 'scalar dumping' of responsibility from national government to local authorities, and then onto local communities, friends' groups and charities in the absence of any national accountability or statutory funding (Smith et al. 2023). The responses within the UK to austerity measures of central government have led to increased heterogeneity as alternative models for revenue extraction and community management are tested; some local governments are better equipped than others to innovate or invest in new structures and processes for managing parks, and place-based sensitivity requires tailored responses (Mell 2019). Whilst transfer of management to charitable trusts in some cases appears to be working to focus attention on a single purpose and diversify opportunities for fund-raising, there is little stringent evidence of their long-term effectiveness, and community management models have been desecrated by the impact of austerity on volunteering capacity. Furthermore, they lack democratic certainty, and are reliant on local power dynamics of those with the social and cultural capital to bring skills and expertise voluntarily, leading to potential exclusion, lack of representation and further entrenched power inequalities (Smith et al. 2023). Even the advocacy documents for community asset transfer urge caution to avoid the uneven treatment of those community groups who express interest through a dedicated strategy for assessing positive and negative impacts on different members of the community (My Community n.d.)

As discussed in Chaps. 4 and 5, participants in the UEP research discussed their motivations for taking part in the management of parks and lobbying for their creation and conservation, and the rewards they gained from civic participation. Volunteering was seen as a good access route to knowing and feeling part of their neighbourhood communities, but was also motivated by the need to fill the gap left by the local authority. In the rural and peri-rural areas in Dartmoor and Scotland, volunteering was a 'moral currency' (Miles and Ebrey 2017, p. 63), with higher levels of participation than in urban neighbourhoods of Manchester and Gateshead, which were characterised by more transient, less affluent populations than their rural counterparts. The difficulties encouraging local people in formal stewardship through friends' group and community gardening projects and the lack of informal care demonstrated by fly-tipping and littering was a source of frustration for the local authorities in Manchester and Salford, particularly in places with transient populations (Gilmore and Lang 2020). There was social stratification of participation found across all case studies, even amongst the 'super-participation' in the Western Isles fostered by community land trust development, where participants commented on their reluctance to take part as 'incomers' and younger people were excluded from positions of power.

If the history of park-making in the nineteenth and twentieth centuries was one of municipalisation, then now it is one of 'de-municipalisation' (Smith et al. 2023). The ideological shift of service provision away from the state, and into the community and private sector mirrors the previous crisis point for municipal public parks in the mid- to late 1990s, as Compulsory Competitive Tendering (CCT) was brought in under the Local Government Act of 1988. Aligned with the ideologies of the Thatcher Conservative government, CCT was rolled out as an efficiency measure compelling local authorities to take their ground maintenance work in all parks to the market. Ironically, as a cost-saving measure this compulsion was potentially counterproductive since even when it was possible for local authorities to show they could offer services more cheaply than private sector counterparts, the regulations of CCT prevented them from applying (Dempsey et al. 2020). Furthermore, the outsourcing of key sector skills stripped many local authorities of their in-house horticultural skills (Dempsey 2023), making marketisation both inevitable and effectively irreversible.

A good practice guide commissioned by the Department of Environment at this time, *People, Parks and Cities* (Greenhalgh and Worpole 1996) undertook consultation research on the advantages and disadvantages of CCT, which effectively split the contracting of maintenance work from that of the security, management and ongoing development of parks. The report identified increased transparency in the costs of keeping parks, and some cost savings, but also an increased need to spend time managing contracts, a neglect of public interest in ensuring staffing and security arrangements are equally supported, the squeezing out of smaller community-based proposals by large bundles of contracts to established firms, and in some cases increased vandalism through the reduction of site-based staff, as the traditional park-keeper was replaced by contracted-in ground staff. At this earlier point of crisis for public parks, budget-cutting and efficiency measures metered out on local authorities were in some degree mitigated by the timely advent of Heritage Lottery Fund to parks funding, who came to the rescue providing more than £850 m over 1996–2016 and creating a turn-of-the-millennium "renaissance" for park conservation (Smith et al. 2023, p. 6). As discussed in Chap. 2, the promotion of public value generated by funding the conservation of parks was seen as ample return on investment, not least since parks are so popular and arguably more legible as heritage objects with local communities, which the HLF strategy ensured were represented alongside local authorities (Gilmore and Lang 2020) through match-funding volunteer value-in-kind within projects (Smith et al. 2023). However, the legacy of privatisation outlasted significant heritage funds for parks, which ended in 2018 as the National Lottery Heritage Fund revised its strategy and ended the Parks for People Programme.

MARKETISATION, ARBORICIDE AND SOCIAL JUSTICE

The impact of contracting out services for environmental management is also being felt outside of park walls and railings. Like many others, the city of Sheffield benefitted from street tree planting, planned into new suburbs for the expanding industrial middle classes from the mid-nineteenth century and incorporated into municipal service delivery and social housing estates for less affluent areas over the centuries (Dempsey 2023). The city had been recognised for its care of green heritage and promotion of community engagement in environmental management in urban public realm. However, the emergence of public-private partnerships (PPP) as governance structures to deliver large-scale capital projects in the 1990s led to the parallel financial mechanism of Private Finance Initiative (PFI), facilitating both private finance and public investment into services to protect public borrowing, delivered by private agencies on long-term contracts. A PFI named *Streets Ahead* was instigated in 2012 in Sheffield as a 25-year programme to manage roadways and highways in the city; it included responsibility for street tree management, won by a private contractor, Amey, who set about replacing large numbers of the city's 36,000 street trees, having identified 75% of them due to the risks posed by their maturity and potential decline (Dempsey 2023, p. 7). Whilst the state of the trees and their need for replacement is moot, the poor management of community relations around tree felling and lack of transparency over how the decisions were made quickly caused an escalation of opposition and protest, not least when it transpired that even as the decisions of the city council and the private contractor were being defended and rationalised in 2017, they had already felled over 5000 trees with an estimated total asset value of over £66 million (Rotherham and Flinders 2019, p. 195).

The loss of trees and the controversy, multiple protests and ill-feeling that surrounded their felling led to a concomitant loss of public trust in their local authority, whose councillors were found to ignore or mislead public interest in a formal inquiry into the 'chainsaw massacre', and who had also encouraged South Yorkshire police to continue arresting protests even when this was seen to be ineffective if not incendiary to the cause (Pidd 2023). It seems the reputational damage fell predominantly on the public side of the PPP which had led to this disastrous strategy, and the private interests were absolved under the ideological shelter of contractual agreements. The removal of street trees and the public protest this causes is not confined to Sheffield, however, has been seen repeatedly in the UK and elsewhere in recent years. Rotherham and Flinders (2019) name this the 'street tree paradox' (p. 196) which mirrors that of the parks value paradox I discuss here and in Chap. 1, and which revolves around the tension between the evidence of the socio-economic value of trees and the desire of local government to govern in a politically attractive manner, performing short-term cost savings by literally cutting down future maintenance costs.

This has been taken to further extremes in Madrid, where in 2022, the City Council, led by Almeida, Madrid's mayor, instigated a programme of street tree felling, filling the remaining holes with concreted-in cobbles so that they cannot be replaced. Reports in the local press document the number of 'cobble pits' at over 4000 in just two years (Casado 2023), significantly impacting the biodiversity and shade coverage of a city known for its extreme temperatures in the middle of a climate emergency (see Fig. 6.1). The felling was carried out by a joint venture of contractors who won the 0.25 m Euro tender, which ironically coincided with the launch of a plan from Almeida's team, in which it promised to plant a tree in each existing empty space in the city. Furthermore, the impact of this new grey cover-up has been felt most in neighbourhoods Puente de Vallecas, Villa de Vallecas and Vicálvaro, working class, left-leaning neighbourhoods who have seemingly received their new cobble pits as a reward to long-term resistance to canvasing from the leading right-wing party. The 'arboricide' of Madrid's trees has been met by public protest, and follows further

Fig. 6.1 Notice in community-planted replacement tree, Lavapies, Madrid. (Photo credit: Sebastian Jones). The sign says in Spanish 'This beautiful Mediterranean hackberry tree was planted by the residents of Lavapies. Look after him and he will give you stupendous shade'

concerns of disemparkment in Madrid, as the city's major park the Retiro suffered closures (Casado 2023). Two further city parks have had felling and removal orders for 1500 trees as construction works for the new metro line were routed into them away from the streets to avoid traffic disruption, triggering mass protests, petitions and further anger at the local and regional authorities (Jones 2023). Communities have taken to guerrilla-planting in their own neighbourhoods, unappeased by the concurrent gesture by the mayor to give every household a plant for their private balcony (Publico 2023).

Whilst there has been media speculation that tree felling and cobble pits in Madrid's poorest districts are acts of political retribution, there is over-whelming evidence that park inequities are often felt in the places that need them most. The Fields in Trust Green Space Index, a methodology for calculating park equity which was updated following the rising signifi-cance of proximate access to green spaces during the pandemic, finds that whilst there is 30 sqm of provision per person in Britain, it is not equitably distributed. The measure suggests that nearly three million people live further than a ten-minute walk from green space, and that the local authorities prioritised within the UK government's levelling-up agenda, which aims to rebalance inter-regional disparities in economic growth and productivity, suffer on average 10% less green space than the national aver-age (Fields in Trust 2022). Fields in Trust had also previously commis-sioned contingent valuation research using government-approved metrics to establish economic proxies for the use and non-use values attributed by individuals to the existence of their local parks. This research, which was triggered by frustrations with the Community and Local Government select committee (CLG 2017) as a further attempt to inform policy, iden-tifies the value of parks as £2.52 per month for the average citizen, a figure that increases to £4.32 for those from lower socio-economic groups, and even more significantly for Black and Minority Ethnic (*sic*) communities to £5.84 per month (Fields in Trust 2018, p. 7).

National participation data also suggests the importance of green spaces to less affluent and more marginalised social groups, with evidence of more frequent visits and a wider range of motivations for visiting parks (Eadson et al. 2021). If such data do not inspire public policy to inter-vene, then this leaves the market to provide. The creation of new green spaces lies in the hands of the developers and would present a significant opportunity for community planning gain, however, the establishment of park spaces near new housing developments has declined in the UK over the last 20 years, as developers protect their profit margins, and with just 6% of parks and green spaces protected legally in England, building on infill green spaces instead (Horton 2022). Parks take up valuable profit space in available land, and their privileging within developments must be seen to equal if not exceed the benefits of building space that can be com-modified more easily, such as office space, entertainment, and retail space, to overcome the public park value paradox. As pressures for more housing construction grow, with the balance of power firmly within the developers' hands, the community need for green space is left unchecked, a victim of de-municipalisation and a further cause of park inequity.

Below, I consider two final examples of parks to explore the tensions around their management as a public good and to foreshadow the ongoing commodification, privatisation and plunder of public space, by returning to the city of Manchester.

CITY CENTRE PARKS FOR ALL

On 22 September 2022, a new city centre park opened in Manchester, the first in 100 years. Mayfield Park is situated next to Piccadilly station adjacent to the Mayfield Depot, a 10,000 square metre building which had previously been a relief passenger station for regional railway lines, a parcels depot and postal sorting office until its closure in 1986. Redevelopment proposals for the cavernous industrial space have included extended station platforms, a coach station and office space for rehoused civil servants, however, it is now a commercial venture, housing cafés, restaurants, bars and 10,000 capacity live music and performance venue owned by Broadwick Venues LTD. The space had already hosted large-scale arts and cultural events, including the dance music club night, the Warehouse Project, and work commissioned by the Manchester International Festival, when it was given this new 'meanwhile' lease of life. It is the home now to Escape to Freight Island, founded and owned by the charmingly named Meat Raffle Ltd., and expanded during the first year of the pandemic with an additional investment of £2m (Houghton 2020). This development took advantage of the opportunity during the pandemic to offer a large, ventilated event space to Manchester audiences, who had suffered a series of additional lockdowns in 2020 and through 2021 as a 'tier three' area subjected to more stringent public health restrictions than elsewhere in the country (Walmsley et al. 2022).

The new park folds out behind this event space, bounded entirely on its longest side by the Mancunian Way, the elevated motorway which forms an inner ring around the city centre. It has four entrances, two each side of the River Medlock that bisects the park, which lead further into Manchester city centre to the west or into the post-industrial landscape of railway arches and underpasses that precede the ward of Ardwick to the east. The park space itself comprises a playground (nearest to the Ardwick side) with giant slides and swings, the rejuvenated riverside with viewing platforms and bridges, variegated landscape design incorporating 140 trees, kingfisher posts, bird boxes, wildflower and wetland planting and lawn areas which incorporate events infrastructure for future festivals (and presumably event overspill from Escape from Freight Island and the Depot). Its promotional materials use the 'green lungs' analogy to describe its role in the city and make much of the park's green and sustainable credentials, pointing out the restoration and re-use of industrial artefacts found during construction, such as Victorian wells, and the remediation of the Medlock, cleared of the urban detritus flowing downstream (Mayfield Development [General Partner] Limited n.d.).

The promotional materials highlight the aspirations of the park space along with its rules and opening hours: dogs on lead, and from dawn to dusk, respectively. They also reveal that there will be a major construction site surrounding the new park over the next ten years, albeit promising that this will not affect the health and safety of park-goers. This is because the park is the first phase of a major mixed-use development led by the developers, U+I, part of the joint venture partnership which includes Manchester City Council, Transport for Greater Manchester and London and Continental Railways. This scheme aims to repurpose and remediate a 24-hectare site next to the railway lines to accommodate 1400 homes, new retail and leisure space and 154,800 sqm of office space in high- and low-rise development on and around the Depot and the park. Delivery is phased and dependent on planning, market demand, commerciality and other external factors, intending mixed grain commercial and residential spaces which are raked around the site, bookended by two 'landmark' high rises to create 'a bowl-like skyline that steps down to celebrate the park at the centre' (Mayfield Partnership 2018, p. 36). As a location, Mayfield has three selling points: firstly, its connection to transport networks including the erstwhile high-speed rail network HS2, now delayed by central government, and to the nearby airport; secondly, its proximity to the Manchester city centre's retail and entertainment attractions, and thirdly, its very own park, with all the associated natural and cultural benefits.

The wider Mayfield scheme has received criticism for the inclusion of two new car parks and the lack of affordable housing in its plans (The Meteor 2020), however, few have criticised the advent of a new city centre park, and its opening was attended in a rare show of unity by Greater Manchester Metro Mayor, Andy Burnham, the leader of Manchester City Council, Bev Craig, and national government Minister for Levelling Up, Housing and Communities, Michael Gove. Mayfield Park is not a municipal public park, however, as the promotional materials also make clear: it is 'a benchmark for privately-owned public space (POPS) [...] it is completely open to the public and free to access' (Mayfield Development (General Partner) Limited n.d.). Scrolling further down the website provides important detail on this benchmark and its implications, as the small print at the bottom reads:

> Mayfield Park is private land. There are no public rights of way over it. The public is however currently permitted by the owner to have non-vehicular access over such routes as the owner may designate from time to time subject to complying with any terms of access that the owner may impose from time to time. (Mayfield Development (General Partner) Limited n.d.)

The Mayfield scheme promises to act as a catalyst for regeneration and growth of the neighbourhoods around it, particularly the area of Ardwick to the southeast, a ward that sits firmly within the highest decile on the Index of Multiple Deprivation (Mayfield Development [General Partner] Limited n.d.). Whilst there are entrances to the park signposted by landmark buildings, it is difficult to see how those living in Ardwick will benefit from the park's existence, and whether they will cross the threshold symbolised by the Mancunian Way and railway lines that currently define the Mayfield neighbourhood. Although the ward does not meet the Fields in Trust Green Space standards for minimum provision, it does have green spaces which are much closer to existing residential areas, including Ardwick Green Park, a site of public assembly in the early nineteenth century and incorporated as a municipal public park in 1867, Gartside Gardens, a former cemetery dating back to 1821, and two community football complexes with sports fields, all within the vicinity of Mayfield. It seemly unlikely that Mayfield Park will attract local communities to participate within its POP space, even with the attractions of its playground, giant slides, bars, and entertainment (Fig. 6.2).

PICCADILLY GARDENS—A TOO-PUBLIC REALM?

Mayfield's city centre green space predecessor is Piccadilly Gardens, a rhomboid zone of a public realm at the heart of the city first landscaped as gardens in the 1930s, and flanked by an existing promenade designed by Joseph Paxton which hosts commemorative statues, walkways and fountains, and a concrete curving feature wall and pavilion. Since its inception as civic public realm, when Sir William Fairbairn planned to create a grand

Fig. 6.2 The heavily branded Mayfield Park and Depot. (Photo credit: Abigail Gilmore)

piazza as a gateway to Manchester's commercial centre (Moore 2004) the site of Piccadilly Gardens has been dogged by controversy. Frequent attempts have been made to find a function and design scheme that solved what is still understood as 'the Piccadilly Question'. The site is situated on land donated by Sir Oswald Mosley, the manorial landowner, for public use in perpetuity in the mid-eighteenth century, which has since been the site for civic institutions, both proposed and realised, including Manchester Royal Infirmary and Lunatic Asylum, from 1755 until relocation in 1908, public baths, and a planned opera house and arts centre. Heavily bombed during the Second World War, the site is surrounded by retail and transport infrastructure, with consequent traffic and fumes.

The concrete wall was designed by Japanese architect Tadao Ando as part of redevelopment in 2003 which also created a pavilion space adjacent and new office buildings at a perpendicular in an attempt to mask the transport interchange and traffic noise to east and north of the Gardens (see Fig. 6.3). Despite these best intentions, both the wall and offices have

Fig. 6.3 "Piccadilly Wall—The North Is Not a Petri Dish". (Photo credit: David Dixon) The title is a reference to the central government imposed lengthy lockdowns and restrictions on hospitality and entertainment specific to Manchester and the North West during the Covid pandemic CC BY-SA 2.0; https://creativecommons.org/licenses/by-sa/2.0/legalcode

been the subject of significant public criticism since they were built, as 'white elephants' and eyesores which dissect the eyelines and desire lines of the gardens, sheltering anti-social behaviour such as drug-taking and littering. The Gardens are frequently used as a meeting point in the city centre; a place for hanging out, an assembly point for protest and for live events and street markets. They are also known for begging, busking, for anti-social behaviour and as a haunt for transients and homeless people. Their position at the end of the street leading to Piccadilly station surrounded by large department stores and chain restaurants, tram and bus routes exposes their users to the pollution and noise of the city. They seem very far aesthetically from the previously sunken gardens depicted in Manchester city centre history books, despite successive consultations, competitions and promises to design out Piccadilly's problems. There is evidence that this space has been problematic from its beginnings: a postcard of the Gardens from the early twentieth century advises an unknown friend:

> I am afraid you would be very much disappointed in Piccadilly. All the white foreground in this photo gives a wrong impression. It is really muddy and dirty and even in fine weather it is not white and clean. The x marks a long row of seats where all the tramps sleep out if the police don't find them first. I see people out of work sat there during the day (Anon, cited in Manchester History Net n.d.).

Despite their proximity to one another and their common claims to provide much-needed city centre green space, the contrast between Mayfield Park and Piccadilly Gardens is stark. Whilst also circumscribed by transport infrastructure, Mayfield has statement biodiversity and design studio aesthetics which hope to claim back the natural habitat from the shopping trolleys on the riverbed and the polluted air of the Mancunian Way. It has sculpted landscaping, variegated planting and features which provide different levels and sightlines. It is closed at night and policed by the security guards who casually signal its private ownership. Meanwhile, Piccadilly Gardens is flat, grimy and mainly concreted, flanked by department stores, bus interchange and office towers, and subject to their commercial gaze. On weekends it is full of ordinary people who come to the city centre to shop, people-watch and hang-out, in some cases to take and deal drugs, beg and sleep. It is public space which acts as an illustration of Di Masso's emplaced practices of urban conflict as 'a central component of public space and a fundamental instrument for achieving the right to the city because it expresses territorially structural power struggles between the accepted publics and the socially unwanted counter-public' (Di Masso 2015, p. 70).

If the embattled and almost too-public space of Piccadilly presents through its agonism the right to the city, the commodified Mayfield, annexed to the hospitality playground of the Depot and soon to be surrounded by mixed use development, also provides 'a text into the city' (Masteron-Algar 2016, p. 83). Treating parks like texts may have the danger of only providing commentary on their surface design, aesthetics architecture, and composition of their landscaping and not the influence of social, political and economic forces. In a critical review of scholarship on Frederick Olmsted's seminal influence on park-making in the United

States and worldwide, Rosenzweig (1984) observes the importance of looking beyond park boundaries for the forces that shape them, bringing together semiotic and aesthetic analyses of parks with multidisciplinary and sociological analysis of their political economy. This is an approach that he takes up in more detail in a later book with Elizabeth Blackmar (Rosenzweig and Blackmar 1992) when studying Central Park through its ages, investigating the elitist public sphere surrounding Olmsted, Mintun and other park advocates in New York's political classes, and exploring the shifting meanings attached to this most famous of public parks in America over time, relating these to broader social, economic and political conditions (Pacyga 1996). Seen through this wide-angle lens, the new Mayfield Park reflects as a text the local structures of feeling of twenty-first-century Manchester, a 'creative city' (Whiting et al. 2022) that hopes to attract the class capital who can appreciate its fine landscaping, respect its rules and provide revenue for its upkeep through their patronage of the Depot. Whilst Piccadilly presents a planning problem, exposing the everyday participation of the 'ordinary city' of Manchester, its homelessness, deviance and drug-taking and also buskers, protestors, multicultures and subcultures, Mayfield is a policy solution, with built-in political capital, anchoring a joint venture PPP which will regenerate a high-land value area and ultimately pay for itself. There was no need, therefore, for the mayor of Manchester to declare at its opening the park was being handed over to the people of the city to look after as their own, as they did in 1846 in Philips Park (Ruff 2016), since this is simply not the case: there is neither right to the park nor the city in this private space.

CONCLUSION

In this book, I bring a multifaceted approach to thinking about public parks to avoid privileging one aspect of their making and management, which may give cause to "rob them of their complexity and to misunderstand their history" (Rosenzweig 1984, p. 290). Unlike Rosenzweig, I have not especially considered Central Park or the profound effects of and influence on Olmsted's park-making, or design thinking in today's parlance, that emerged from his visits to Birkenhead in the northwest of England in 1850. However, following his lead I recognise the value of looking beyond parks' physical boundaries in the framing I have chosen for my examination of predominantly English parks. I have treated parks as texts for the cities and towns of their location, spaces, and lenses through which to examine power relations and policy logics in the cultural domain.

They are examined as sites of cultural and everyday participation and locational citizenship, places to test regulatory practices in strategies for moral improvement and urban reform, and also spaces for memory-making and social encounter, for aesthetic and sensory experiences of natural heritage, the articulation of cultural values and expression of identity.

Public parks bring together the *cité* and the *ville* by making public space for open-form serendipity and synchronicity (Sennett 2018) and mediated conviviality (Barker 2017; Barker et al. 2019); as common-pool properties, they offer generative values through their contribution to well-being and healthy cities, but also the opportunity for extraction and rivalry through commodification and enclosure. Examining municipal parks reveals a policy paradox, as those originally responsible for their ownership in public trust are under pressure from austerity, a retracting central government, and the interests of the market, forcing a turn to communities for their stewardship and preservation. De-municipalisation is a process rather than an outcome, and symptomatic of these wider malaises.

The consideration of the right to the park as a right to the city gives us pause to think about how planners, planning and policy can support cultural democracy and the practising of public space. Those involved in making parks form a cultural public sphere which is narrow and elitist historically, however, well-intended to serve and make a culturally literate, morally reformed and economically productive public body. As the right to the park falls away from the public and is enclosed by the private, so we may see further erosion of the right to the city, following Lefebvre (1968) and Harvey (2020) as an ouvre where all citizens participate (Mitchell 2003). The significant values and meanings that local public parks have to their everyday participants, in mundane times and during periods of trauma and crisis, such as the COVID-19 pandemic, are the basis for individual and communal planning, emplaced practice and common-pool stewardship of the right to the city. Through examining the participation narratives of those different constituencies in their making and practising, I have argued that parks are spaces for cultural policy in their regulation of the public body, and also in their promotion of agonism, struggle and the resistance of counter-publics, as sites of participation that generate cultural values, expressions and identities, mediating and facilitate common experiences and co-existence and, therefore, so much more than simply green lungs or commodifiable assets for place-making. As a fundamental part of culturally democratic place governance, they deserve the attention of the government to restore public trust and retain their significant value in everyday life.

REFERENCES

Barker, A. 2017. Mediated conviviality and the urban social order. *British Journal of Criminology* 57: 848–866.

Barker, A., A. Crawford, N. Booth, and D. Churchill. 2019. Everyday encounters with difference in urban parks: Forging 'openness to otherness' in segmenting cities. *International Journal of Law in Context* 15 (4): 495–514. https://doi. org/10.1017/S1744552319000387.

Belfiore, E. 2020. Whose cultural value? Representation, power and creative industries. *International Journal of Cultural Policy* 26 (3): 383–397. https:// doi.org/10.1080/10286632.2018.1495713.

———. 2022. Is it really about the evidence? Argument, persuasion, and the power of ideas in cultural policy. *Cultural Trends* 31 (4): 293–310. https:// doi.org/10.1080/09548963.2021.1991230.

Bell, D., and K. Oakley. 2014. *Cultural policy.* London: Routledge.

Cairney, P. 2016. *The politics of evidence-based policy making.* New York: Springer.

Casado, D. 2023. Almeida has closed 2,188 tree pits in Madrid and plans to close another 2,044 this year. *El Diario.* https://www.eldiario.es/madrid/somos/almeida-clausurado-2-188-alcorques-madrid-preve-cerrar-2-044-ano_1_9938633.html. Accessed 10 May 2023.

CLG. 2017. Communities and local government committee public parks inquiry. https://publications.parliament.uk/pa/cm201617/cmselect/cmcomloc/45/45.pdf. Accessed 19 Feb 2022.

Crowe, L. 2018. The future of public parks in England: Policy tensions in funding, management and governance. *People, Place and Policy* 12 (2): 58–71.

Dempsey, N. 2023. The Sheffield street tree dispute: A case of "business as usual" urban management? *Journal of Environmental Planning and Management.* https://doi.org/10.1080/09640568.2023.2201965.

Dempsey, N., M. Burton, and J. Selin. 2020. In-house, contracted out… or something else? Parks and road management in England. In *Marketization in local government: Diffusion and evolution in Scandinavia and England,* ed. Andrej Christian Lindholst and Morten Balle Hansen, 101–116. London: Palgrave Macmillan.

Di Masso, A. 2015. Micropolitics of public space: On the contested limits of citizenship as a locational practice. *Journal of Social and Political Psychology* 3 (2): 63–83. https://doi.org/10.5964/jspp.v3i2.322.

Eadson, W., C. Harris, S. Parkes, B. Speake, J. Dobson, and N. Dempsey. 2021. *Why should we invest in parks? Evidence from the Parks for People programme.* NLHF. https://www.heritagefund.org.uk/about/insight/evaluation/parks-people-why-should-we-invest-parks. Accessed 21 May 2023.

Fields in Trust. 2018. *Revaluing our parks and green spaces.* London: Fields in Trust.

———. 2022. Policy: Green space index reveals importance of local parks for achieving levelling-up. *Fields in Trust,* May 18. https://www.fieldsintrust.org/News/green-space-index-reveals-importance-of-local-parks-for-achieving-levelling-up. Accessed 13 May 2023.

Gilmore, A., and L. Lang. 2020. Talking, walking and making in Cheetham Park. *Conjunctions* 7 (2): 1–20. https://doi.org/10.7146/tjcp.v7i2.119258.

Gray, C. 2002. Local government and the arts. *Local Government Studies* 28 (1): 77–90.

Greenhalgh, L., and K. Worpole. 1996. *People, parks and cities: A guide to current good practice in urban parks,* A report for the Department of Environment. London: HMSO.

Gurian, E.H. 2005. *Civilizing the museum: The collected writings of Elaine Heumann Gurian.* London: Taylor & Francis.

Harney, S., and F. Moten. 2013. *The undercommons: Fugitive planning and black study.* Research Collection Lee Kong Chian School of Business, 1–165.

Harvey, D. 2003. The right to the city. *International Journal of Urban and Regional Research* 27 (4): 939–941.

———. 2020. The right to the city: New left review (2008). In *The city reader,* ed. Richard T. LeGates and Frederic Stout. 281–289. London: Routledge.

Horton, H. 2022. Parks near new homes shrink 40% as developers say they cannot afford them. *The Guardian,* May 3. https://www.theguardian.com/cities/2022/may/03/green-space-decline-housing-developer-england-wales-plead-poverty-research-finds. Accessed 05 May 2023.

Houghton, T. 2020. Escape to Freight Island announces fresh £2m investment to develop and expand exciting Manchester venue. *Business Live,* November 24. https://www.business-live.co.uk/economic-development/escape-freight-island-announces-fresh-19334440. Accessed 7 May 2023.

Institute for Government. 2020. Explainer: Local government funding in England. https://www.instituteforgovernment.org.uk/explainer/local-government-funding-england#:~:text=Local%20authority%20'spending%20power'%20%E2%80%93,fallen%20by%2016%25%20since%202010. Accessed 18 Apr 2023.

Jones, S. 2023. Thousands to protest in Madrid over 'barbaric' plan to fell over 1,000 trees. *The Guardian.* Available from: https://www.theguardian.com/world/2023/feb/17/madrid-protest-plan-cut-down-park-trees-arganzuela. Accessed 17 May 2023.

Kaszynska, P., and G. Crossick. 2016. *Understanding the value of arts and culture, Project Report.* Swindon: AHRC.

Lefebvre, H. 1968. *Le Droit à la ville.* Paris: Anthropos.

Manchester History Net. n.d. Piccadilly Gardens. https://manchesterhistory.net/manchester/squares/piccadillygardens.html. Accessed 08 May 2023.

Martinsson, K., D. Gayle, and N. McIntyre. 2022. Funding for England's parks down £330m a year in real terms since 2010. *The Guardian.* https://www.theguardian.com/environment/2022/aug/23/funding-for-englands-parks-down-330m-a-year-in-real-terms-since-2010. Accessed 04 May 2023.

Masteron-Algar, A. 2016. *Ecuadorians in Madrid: Migrants' place in urban history.* New York: Palgrave Macmillan.

Mayfield Development (General Partner) Limited. n.d. Mayfield Park: A place for all. https://mayfieldpark.com/. Accessed 05 May 2023.

Mayfield Partnership. 2018. Mayfield regeneration framework, May. https://www.manchester.gov.uk/downloads/download/6851/mayfield_srf_february_2018. Accessed 08 May 2023.

McGuigan, J. 2004. *Rethinking cultural policy.* Maidenhead: Open University Press.

Mell, I. 2019. Beyond the peace lines: Conceptualising representations of parks as inclusionary spaces in Belfast, Northern Ireland. *Town Planning Review* 90 (2): 195–218. https://doi.org/10.3828/tpr.2019.13.

Miles, A., and J. Ebrey. 2017. The village in the city: Participation and cultural value on the urban periphery. *Cultural Trends* 26 (1): 58–69. https://doi.org/10.1080/09548963.2017.1274360.

Mitchell, D. 2003. *The right to the city: Social justice and the fight for public space.* New York: Guilford Press.

Moore, J. R. 2004. Urban space and civic identity in Manchester 1780–1914: Piccadilly Square and the art gallery question. *The Historic Society of Lancashire & Cheshire* 153: 87–123.

My Community. n.d. Parks and green spaces: Community asset transfer. https://mycommunity.org.uk/files/downloads/Parks-and-Green-Spaces-Community-Asset-Transfer-B.pdf. Accessed 13 May 2023.

Ostrom, E. 1990. *Governing the commons: The evolution of institutions for collective action.* New York: Cambridge University Press.

Pacyga, D. 1996. Central Park and the public sphere. *Journal of Urban History* 22 (5): 649–655. https://doi-org.manchester.idm.oclc.org/10.1177/009614429602200505.

Pidd, H. 2023. Sheffield city council behaved dishonestly in street trees row, inquiry finds. *The Guardian.* https://www.theguardian.com/uk-news/2023/mar/06/sheffield-city-council-behaved-dishonestly-in-street-trees-row-inquiry-finds. Accessed 13 May 2023.

Publico. 2023. Que cada balcon de Madrid tenga una planta los tuiteros alucinan con la ultima receta ambientalista de ayuso. *Publico*, May 17. https://www.publico.es/tremending/2023/05/17/que-cada-balcon-de-madrid-tenga-una-planta-los-tuiteros-alucinan-con-la-ultima-receta-ambientalista-de-ayuso/. Accessed 29 May 2023.

Rosenzweig, R. 1984. Review: The parks and the people: Social history and urban parks. *Journal of Social History* 18 (2): 289–295. Oxford University Press, Oxford. Stable URL: https://www.jstor.org/stable/3787289.

Rosenzweig, R., and E. Blackmar. 1992. *The park and the people: A history of Central Park.* Ithaca/London: Cornell University.

Rotherham, I., and M. Flinders. 2019. No stump city: The contestation and politics of urban street-trees—A case study of Sheffield. *People, Place and Policy* 12 (3): 188–203. https://doi.org/10.3351/ppp.2019.8283649746.

Sennett, R. 2018. *Building and dwelling: Ethics for the city.* London: Allen Lane.

Smith, A. 2019. Event takeover?: The commercialisation of London's parks. In *Destination London: The expansion of the visitor economy,* ed. Andrew Smith and Anne Graham, 205–224. London: University of Westminster Press.

———. 2021. Sustaining municipal parks in an era of neoliberal austerity: The contested commercialisation of Gunnersbury Park. *Environment and Planning A: Economy and Space* 53 (4): 704–722. https://doi.org/10.1177/0308518X20951814.

Smith, A., M. Whitten, and M. Ernwein. 2023. De-municipalisation? Legacies of austerity for England's urban parks. *The Geographical Journal,* 1–12. https://doi.org/10.1111/geoj.12518.

Standing, G. 2019. *Plunder of the commons: A manifesto for sharing public wealth.* London: Penguin Random House.

The Guardian. 2023. The Guardian view on parks: An asset that should be for everyone. Editorial. https://www.theguardian.com/commentisfree/2023/apr/23/the-guardian-view-on-parks-an-asset-that-should-be-for-everyone. Accessed 07 May 2023.

The Meteor. 2020. Does the Mayfield proposal offer the radical change Manchester desperately needs or will the city be left unsatisfied? *The Meteor,* September 1. https://themeteor.org/2020/09/01/does-the-mayfield-proposal-offer-the-radical-change-manchester-desperately-needs-or-will-the-city-be-left-unsatisfied/. Accessed 07 May 2023.

Walmsley, B., A. Gilmore, D. O'Brien, and A. Torrigiani. 2022. *Culture in crisis: Impacts of Covid-19 on the UK cultural sector and where we go from here.* Leeds: Centre for Cultural Value.

Whiting, S., T. Barnett, and J. O'Connor. 2022. 'Creative city' R.I.P.? *M/C Journal* 25 (3) https://doi.org/10.5204/mcj.2901.

Williams, R. 1983. *Keywords: A vocabulary of culture and society.* Oxford: Oxford University Press.

Index[1]

[1] Note: Page numbers followed by 'n' refer to notes.

© The Author(s), under exclusive license to Springer Nature Switzerland AG 2023
A. Gilmore, *Culture, Participation and Policy in the Municipal Public Park*, Palgrave Studies in Cultural Participation, https://doi.org/10.1007/978-3-031-44277-3